Lecture Notes in Artificial In

Edited by J. G. Carbonell and J. Siekma

T0238249

Subseries of Lecture Notes in Computer Science

Francisco Botana   Tomas Recio (Eds.)

Automated Deduction in Geometry

6th International Workshop, ADG 2006
Pontevedra, Spain, August 31-September 2, 2006
Revised Papers

 Springer

Series Editors

Jaime G. Carbonell, Carnegie Mellon University, Pittsburgh, PA, USA
Jörg Siekmann, University of Saarland, Saarbrücken, Germany

Volume Editors

Francisco Botana
EUET Forestal
Universidad de Vigo
Campus A Xunqueira
36005 Pontevedra, Spain
E-mail: fbotana@uvigo.es

Tomas Recio
Universidad de Cantabria
Departamento de Matemáticas, Estadística y Computación
Avenida de los Castros, s/n, 39071 Santander, Spain
E-mail: tomas.recio@unican.es

Library of Congress Control Number: 2007941260

CR Subject Classification (1998): I.2.3, I.3.5, F.4.1, I.5, G.2

LNCS Sublibrary: SL 7 – Artificial Intelligence

ISSN   0302-9743
ISBN-10   3-540-77355-X Springer Berlin Heidelberg New York
ISBN-13   978-3-540-77355-9 Springer Berlin Heidelberg New York

Springer is a part of Springer Science+Business Media

springer.com

© Springer-Verlag Berlin Heidelberg 2007
Printed in Germany

Typesetting: Camera-ready by author, data conversion by Scientific Publishing Services, Chennai, India
Printed on acid-free paper   SPIN: 12206292   06/3180   5 4 3 2 1 0

Preface

After five successful editions (Toulouse, 1996; Beijing, 1998; Zurich, 2000; Linz, 2002, Gainesville, Fl., 2004), the series of international workshops on Automated Deduction in Geometry (ADG) has consolidated its fundamental role concerning the scientific community working on the interaction between geometry and automated deduction. From August 31 to September 2, 2006, a new ADG meeting took place at the Pontevedra (Galicia, Spain) campus of the University of Vigo, as a satellite event of the International Congress of Mathematicians (Madrid, August 22–30, 2006). We acknowledge the financial support for ADG 2006, provided by the University of Vigo and the Spanish Ministerio de Educación y Ciencia under grant MTM2005-24580-E.

It was a fruitful meeting – made possible by the work of the Organizing Committee (see next page) – for exchanging ideas and for the presentation of original results and software novelties – 21 contributions in total – under the scientific guidance of the Program Committee (listed on the next page). Moreover, it was a privilege to receive the lectures of our distinguished guest speakers, Thomas Hales (U. Pittsburgh) and Martin Peternell (T.U. Wien), dealing with the so-called Flyspeck project, i.e., the automatization of Hales' solution to Kepler's conjecture, and with rational offset surfaces and related issues in CAGD, respectively.

Shortly after the meeting, a call for papers – within the scope of ADG, but with content not necessarily related to a presentation at ADG 2006 – was launched. After a long and detailed process of peer review and revision, we – the editors – have selected the 13 papers of this volume, as a testimony of the current state of the art concerning automated deduction in geometry.

This volume includes a paper by X. Chen and D. Wang proposing a system in the form of a textbook – an electronic geometry textbook, to be more precise – for managing geometric knowledge dynamically, effectively, and interactively. The contribution by T. Hales, in the context of the "Flyspeck" project, describes an algorithm that decides whether a region in three dimensions, described by quadratic constraints, is equidecomposable with a collection of primitive regions and, when a decomposition exists, finds the volume of the given region. P. Janičić and P. Quaresma present an application of automatic theorem proving in the verification of constructions made with dynamic geometry software. The paper by P. Lebmeir and J. Richter-Gebert proposes an algorithm for automated recognition of computationally constructed curves and discusses several aspects of the recognition problem. R. H. Lewis and E. Coutsias deal with polynomial systems, flexibility of three-dimensional objects, computational chemistry, and computer algebra. D. Lichtblau's contribution on computational real enumerative geometry discusses the number and reality of the cylinders generically determined by five points in R^3. D. Michelucci and S. Foufou address the detection of dependences in geometric constraints solving, and propose to use the recently published witness method. The paper by A. Montes and T. Recio merges two techniques (automatic discovery and minimal canonical comprehensive Gröbner systems) to discover missing hypotheses in generally false statements. J. Narboux describes the mechanization of the proofs of the first eight chapters of the classic book "Metamathematische Methoden in der Geometrie" by Schwabäuser, Szmielew and Tarski. The paper by P. Pech deals with the problem, posed by Chou long ago, of finding a natural geometry problem where hypotheses are not described by a radical ideal, such as the existence of regular polygons (pentagons, heptagons) of even dimension. E. Roanes–Macías and E. Roanes–Lozano present a Maple package, on the interaction of computer algebra and dynamic geometry, for investigating problems about configuration theorems in 3D geometry and performing mechanical theorem proving and discovery. P. Todd presents an interactive symbolic geometry package, "Geometry Expressions," generating algebraic formulas from geometry in an interactive style which is convenient not only for high school students, but also for mechanical engineers, graphics programmers, architects, surveyors, machinists, and many more. Finally, the paper by L. Yang and Z. Zeng, employing a method of distance geometry, achieves the symbolic solution to the following problem: express the edge-lengths of a tetrahedron in terms of its heights and widths.

We, the editors, would like to thank the efforts of so many anonymous referees involved in the process of selection and improvement of the submitted papers. We think that, as a consequence of their work, this collection of papers, although necessarily incomplete, shows the lively variety of topics and methods and the current applicability of ADG to different branches of mathematics and to other sciences and technologies.

October 2007   Francisco Botana
                                                                      Tomas Recio

Organization

Invited Speakers

Thomas Hales (University of Pittsburgh, USA)
Martin Peternell (Vienna University of Technology, Austria)

Organizing Committee

Francisco Botana (Pontevedra, Spain), Chair
Miguel Abánades (Madrid, Spain)
Jesús Escribano (Madrid, Spain)
Eugenio Roanes–Lozano (Madrid, Spain)
José L. Valcarce (Santiago, Spain)

Program Committee

Tomas Recio (Santander, Spain), Chair
Hirokazu Anai (Kawasaki, Japan)
Giuseppa Carrà Ferro (Catania, Italy)
Shang-Ching Chou (Wichita, USA)
Arthur Chtcherba (New York, USA)
Luis Fariñas del Cerro (Toulouse, France)
Jacques D. Fleuriot (Edinburgh, UK)
Xiao-Shan Gao (Beijing, China)
Laureano González-Vega (Santander, Spain)
Hoon Hong (Raleigh, USA)
Deepak Kapur (Albuquerque, USA)
Hongbo Li (Beijing, China)
Manfred Minimair (South Orange, USA)
Jürgen Richter-Gebert (Munich, Germany)
Meera Sitharam (Gainesville, USA)
Thomas Sturm (Passau, Germany)
Quoc-Nam Tran (Beaumont, USA)
Dongming Wang (Beijing,China/Paris, France)
Neil White (Gainesville, USA)
Franz Winkler (Linz, Austria)
Lu Yang (Chengdu, China)
Zhenbing Zeng (Shanghai, China)

Table of Contents

Towards an Electronic Geometry Textbook

Xiaoyu Chen[1] and Dongming Wang[1,2]

[1] LMIB – School of Science, Beihang University, Beijing 100083, China
[2] Laboratoire d'Informatique de Paris 6, Université Pierre et Marie Curie – CNRS
104 avenue du Président Kennedy, F-75016 Paris, France

Abstract. This paper proposes a system in the form of a textbook
for managing geometric knowledge dynamically, effectively, and interactively. Such a system, called an *Electronic Geometry Textbook*, can be
viewed or printed as a traditional textbook and run as dynamic software
on computer. The knowledge in the textbook is being formalized by using
standard formal languages and may be processed by software modules
developed for geometric computing and reasoning, diagram generation,
and visualization. The textbook can be generated automatically by organizing and presenting the textbook data according to some specifications.
The system allows the user to manipulate (query, modify, restructure,
etc.) the textbook with automated consistency checking. We present the
main ideas on the design of the electronic geometry textbook, explain
the features of the system, propose five phases of creating and managing
the geometric knowledge in the textbook, discuss the involved tasks and
some of the fundamental research problems in each phase, and report
our progress and experiments on a preliminary implementation of the
system.

1 Motivation and Introduction

Elementary geometry has been developed for more than 2000 years and an enormous amount of geometric knowledge has been accumulated. It is now desirable
to digitalize, organize, and process various kinds of geometric knowledge including concepts, objects, axioms, theorems, diagrams, texts, and computing and
reasoning mechanisms using advanced computer technology to make them more
easily accessible, manageable, and usable.

In recent decades, remarkable progress has been made on computer-aided geometric problem solving. Most of the theorems in elementary geometry can now
be proved or even rediscovered automatically on computer. Software systems
are available for generating algebraic and readable proofs and drawing dynamic
diagrams automatically. However, research in this direction is by no means ex-
hausted and there is an urgent need to exchange data among different dynamic
geometry software systems (such as Cinderella [5], GCLC [7], and GEOTHER [15])
and to integrate them together with other mathematical software like computer
algebra systems and automated theorem provers to extend their power and functionality. Little work has been done on the management of geometric knowledge

F. Botana and T. Recio (Eds.): ADG 2006, LNAI 4869, pp. 1–23, 2007.

and there is no standard geometric knowledge base that can be accessed and used by different geometric software systems. Although GeoThms [13] integrates GCLC with a repository of theorem statements, their proofs (generated automatically), and the corresponding illustrations in a web interface, relations among the geometric problems in the database have not been considered.

When speaking about managing knowledge, we may first think about textbooks where knowledge is represented systematically and hierarchically according to certain logical structures, e.g., from the simplest to the most complicated and from the basic to the advanced. Due to the well-organized structure of the domain knowledge, textbooks play an important role in education and research to store, manage, and impart knowledge for new learners. Therefore, constructing dynamic, interactive, and machine-processable textbooks (instead of traditional static textbooks) is an interesting project of research and development. The Electronic Geometry Textbook (EGT) aims at providing such a tool by integrating elementary plane geometric knowledge and software modules into a single computing environment to support geometry education, research, and application. The idea of developing such an integrated software system in the form of a textbook for systematical and efficient management of geometric knowledge originates from the second author who has been working on automated geometric reasoning in the last two decades. The first author has been stimulated to elaborate the idea and to carry out the actual implementation. The start of this project has also been motivated considerably by the work of Zeilberger [17] and his invited talk at ADG 2004.

The purpose of this paper is to propose a new style of knowledge management and to present the design and preliminary implementation of a software system in the form of a dynamic textbook for managing geometric knowledge interactively: the system can be viewed or printed as a traditional textbook and can run as dynamic software on computer. The knowledge in the textbook, being formalized by using standard formal languages, can be processed by software modules developed for geometric computing and reasoning, diagram generation, and visualization and can be enriched systematically by "self-learning". The textbook may be generated automatically by organizing and presenting the textbook data according to the user's specifications. The system allows the user to manipulate (search, modify, restructure, etc.) the textbook with automated consistency checking (of its logical structure). We will present the main ideas on the design of the electronic geometry textbook, explain the features of the system, propose five phases of creating and managing the geometric knowledge in the textbook, discuss the involved tasks and some of the fundamental research problems in each phase, and report our progress and experiments on a preliminary implementation of the system. As the implementation is still at an early stage, our emphasis here is placed mainly on the design methodology.

While plane Euclidean geometry is the target of our current investigation, the basic principles and ideas discussed in this paper apply to any geometry or even any subject of mathematics.

2   Objective and Design Methodology

In this section, we provide a short review on the state of the art of mathematical knowledge management, describe the main objective of our EGT project, and discuss our methodology for the EGT database and system design.

2.1   Mathematical Knowledge Management

Currently mathematical knowledge is archived and stored mainly in printed and/or electronic documents, like books and articles. It is presented in a static way and searching for items in such documents can be done only syntactically. The structure of the documents cannot be changed and the knowledge therein cannot be processed by problem solvers, theorem provers, or symbolic calculators.

As Internet has become a major channel for information service, various efforts have been devoted to making mathematical knowledge available on the Internet and exchangeable among different software programs. Some markup languages such as MathML [16] have been developed, making it possible to display and communicate mathematical formulas on the web. As an application of XML, MathML benefits from the tools existing for XML file manipulation. Although it does offer some semantics for symbols in mathematical formulas, the set of symbols supported, when compared to those used by working mathematicians, is very restricted. To ameliorate this situation, projects like OpenMath [4] and OMDoc [8] have emerged. The OpenMath standard focuses on describing the semantic meanings of mathematical objects instead of their appearance by using *content dictionaries*, in which mathematical symbols are defined syntactically and semantically and thus allowed to be exchanged between computer programs. Content dictionaries can be stored in databases or published on the world-wide web. OMDoc is an extension to the OpenMath standard by markup for the structure and the theory level of mathematical documents, adding capabilities of describing the mathematical context of the used OpenMath objects. These languages make mathematical knowledge not only machine-readable but also machine-understandable and provide a foundation for developing, communicating, and publishing mathematics on the web.

Mathematical software systems for symbolic and numeric computation, formal reasoning, proof checking, algorithm verification, etc. need domain knowledge to support relevant (automated) activities. Some systems, such as Mizar [14] whose objective focuses on writing and checking formal mathematics written in the Mizar language, provide mathematical knowledge bases (that contain a large amount of domain knowledge in different mathematical theories) and facilities to browse formal mathematical documents. Other systems like Theorema [12] provide environments for building mathematical knowledge bases in a systematical way and such bases can be browsed, extended, and used for teaching, learning, and mathematical discovering. Although mathematical contents embedded in such documents can be processed inside the systems, the documents lack the characteristics of traditional textbooks: mathematical contents are not presented in natural style, the structures of the mathematical documents are static and passive, the logical relations of knowledge are not considered as in traditional textbooks, and no tool is provided to construct, manipulate, and present these documents effectively.

There are few facilities for data exchange among the systems mentioned above and it is very difficult to share, reuse, and interact on domain knowledge resources. Repeated developments are waste of efforts, time, and energy. To integrate system resources, some mathematical knowledge engineering projects, including MBase [9], HELM [3], the Formal Digital Library [2], and the NIST Digital Library of Mathematical Functions [11], have emerged, aiming at building general mathematical knowledge bases for retrieving, representing, acquiring, and reusing various kinds of mathematical domain knowledge on the web. As the knowledge in the bases is represented at the sematic level, it is possible to make the knowledge bases serve for different levels of need, e.g., automated theorem proving, automated programming, and mathematical education. Some projects make use of web-based semantic knowledge bases to develop mathematical intelligent education environments such as ActiveMath [1] which is designed for students to proceed with e-learning. The courseware (textbook) is produced automatically by weighing the student's preferences. Different preferences may lead to different textbook configurations, but manipulating (such as reconstructing or modifying) the textbook is not allowed.

EGT will synthesize the functionality of document creation and manipulation together with automated processing of mathematical contents in the document. Similar to ActiveMath, EGT is also based on a formalized knowledge base in which mathematical contents can be easily converted to the internal representations of other application software. However, EGT is designed mainly as a tool for human authors to construct their own textbooks with automated verification of structural consistency. The process of producing the textbook is mostly human-driven and manipulations on the textbook are allowed and may lead to new, modified, or improved versions of the textbook.

2.2 Objective and System Description

As one of the most fundamental and oldest subjects of mathematics, geometry is founded on graphical objects abstracted from the real visual world. Geometric computation, reasoning, and visualization require the support and integration of logical deduction mechanisms, effective algebraic methods, and graphical drawing tools, involving both abstract quantities and intuitive figures. Moreover, there is no other mathematical subject than geometry in which there are so many highly interesting and fascinating theorems and such theorems may be proved automatically on computer. In fact, automated deduction is much more mature and successful in geometry than in any other domain of mathematics. When mathematical knowledge is organized in a textbook, it is important to track the logical clues of the knowledge, such as how a concept is introduced and how a theorem is used in the proof of another theorem. The availability of automated reasoning devices is a prerequisite for knowledge organization.

Therefore, we consider geometry a unique and rich subject of mathematics that should be chosen for study in the context of knowledge management. In this study, the full power of computer for symbolic, numeric, and graphical computing and data processing may be used and our idea and methodology may be effectively tested.

The objective of our textbook project is to provide a tool for human authors to construct dynamic, interactive textbooks (see Fig. 1).

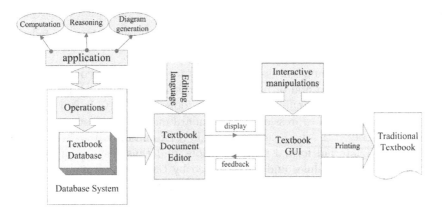

Fig. 1. System Framework

- The formulation and representation of geometric knowledge and the structure of the textbook should be *standardized*, so that the knowledge, when digitalized, can be processed with little extra effort by different software modules existing or being developed for geometric computing and reasoning, interactive or automated dynamic diagram generation, and visualization of geometric objects and the results produced by those modules can be easily integrated into and displayed in the textbook.
- Under this standardization, a *textbook database* will be built, collecting textbook data which refer to as formalized geometric knowledge and related contents (such as literature information, background of remarkable theorems, and explanation texts), and new knowledge may be incorporated properly and automatically into the textbook database by implementing a self-learning and extracting mechanism. The purpose of building such a textbook database is to provide a dataset for reusing and sharing textbook data and semantic information for various applications.
- The user will be able to *edit* his/her own *textbook documents* in an editing language of "mixture" style.
- A *dynamic* textbook graphical user interface (GUI) can be created automatically to display the textbook document by translating formalized geometric statements into natural languages in traditional style automatically.

- The user is permitted to *manipulate* (such as search, modify, and restructure) the textbook or parts thereof and process the knowledge (such as proving theorems, performing computations, and drawing diagrams) in a visual environment using available software modules.
- The system will *verify* some geometric statements by employing other available software modules and *check* some structural consistency of the constructed document in the sense of textbook.

2.3   Design Methodology

The creation and management of the textbook can be divided into the following five phases.

Classification. Mathematical documents contain two kinds of contents, mathematical expressions and ordinary texts, which are mixed and not separated in commonly used document processing tools like LATEX. In our EGT system, mathematical expressions and standard mathematical statements such as "let ... be ...", "if ... then ...", and "... if and only if ..." are formalized (with internal formal representations and thus may be processed by other software modules and manipulated by the user). Ordinary (explanation) texts are clearly identified. We standardize the structure of textbooks by classifying textbook contents with hierarchy and provide the user with standard *templates* for uploading the contents. The classification of geometric objects, concepts, and knowledge "segments" is also a prerequisite for the formalization of geometric knowledge. We shall discuss the classification of textbook contents and geometric knowledge in Section 3.1.

Organization. After textbook contents are classified into parts, our problem is to assemble the chosen "parts" in a reasonable "order" for a concrete textbook. Although our EGT system is human-driven (i.e., how to organize the textbook is decided by the user), the system is capable of verifying the suitability and consistency of the logical structure of the textbook according to some characteristic rules on the structure of standard textbooks and providing the user with suggestions when such rules are violated. The reader will find more details in Section 3.2.

Representation. The above two phases are mainly concerned about the macro-structural aspects of the textbook. As for the contents of each "template", two types of information need be provided: the information for displaying the contents in natural style and the underlying formal representation for interacting with other modules. How to define the two types of information, in particular the semantic formal representation of knowledge and the information needed for various applications? Several data definition models will be described in Section 3.3.

Manipulation. Based on the classification and representation, a language for editing textbook documents has to be developed. How to present formal data in natural style and how a user interacts with the GUI? We propose a kind of language in a mixture style to edit textbook documents with possible operations on the GUI and a "pattern-substitution method" for translating formal statements into natural languages for display (see Section 4.1).

Computation/deduction. To make the system dynamic, its interaction with other software modules is necessary. The main question is how to convert the formal statements of EGT into the internal representations used in other application software. The conversion need be based on semantic information, i.e., the formal definitions of the concepts used in the statements. We shall use an example to illustrate the idea of how to ease the process of conversion in Section 4.2.

In the following two sections, we will explain these phases in more detail.

3   Textbook Database Creation

3.1   Classification

To standardize the structure of textbooks, we need some terminology. By a *segment* we mean a minimal unit of knowledge in the database or a textbook. A segment cannot be broken during data and knowledge processing. One knowledge segment may be surrounded by other knowledge segments. We use *block* to represent the aggregation of these interrelated segments. Figure 2 shows how a traditional textbook can be created by structuring its contents into segments, blocks, sections, and chapters.

In order to construct, manipulate, and manage segments uniformly, we need to classify them into different *classes*, which abstract the sets of segments with common properties. A *template* represents the definition of a class which standardizes the needed data. Every segment in the class can be considered as an *instance* in the sense of programming and can be constructed according to the corresponding template.

For example, the segments in a textbook may be divided into two families of classes: specification classes and knowledge classes. Specification classes include

– Note: describing the background or explaining the meanings of certain knowledge;
– Example: illustrating certain knowledge;
– Exercise: providing problems related to certain knowledge;
– Graph: visualizing certain knowledge;
– Proof or Calculation: demonstrating theorems, lemmas, formulas, etc.

Knowledge classes cover all the knowledge segments and include Definition class, Axiom class, Lemma class, Theorem class, Corollary class, Property class, and Formula class. For instance, the definition of a triangle may be considered as an object of the Definition class at the level of knowledge segment.

Furthermore, in order to formalize knowledge, we need finer classifications, e.g., to specify logical relations among knowledge statements, at the semantic level. We use *statements* to represent the contents in each segment.

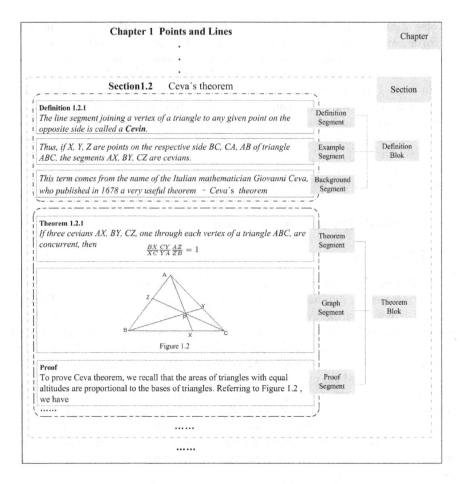

Fig. 2. Structuring the Textbook

A definition defines a new concept using already defined ones. Other knowledge statements are made up of concepts. Some statements like theorems and formulas are used to describe properties about and logical relations (such as "implication" and "equivalence") among concepts. We use Concept class to abstract all the concepts. Geometric concepts may be classified, for instance, into four types

- *geometric objects*: point, line, triangle, orthocenter of a triangle, circle, etc.
- *object relations*: parallelism, intersection, similarity, etc.
- *geometric quantities*: length of a segment, area of a circle, ratio of two segments, etc.
- *quantity relations*: equal to, less than, larger than, etc.

Every concept has a formal definition. According to the statements of definitions, geometric concepts may be divided into two kinds:

- *basic concepts* (including basic geometric objects like point and line, object relations like parallelism, geometric quantities like the length of a segment, and quantity relations like equal to), whose definitions are stated by means of natural language or diagrams, and
- *derived concepts* (such as the circumcenter of a triangle and the orthic triangle of a triangle), whose definitions can be formalized by using basic concepts.

For example, we may consider point, segment, ray, line, angle, triangle, and circle as basic concepts in plane Euclidean geometry. Then many other geometric concepts (such as the circumcenter and the median of a triangle and the diameter of a circle) may be considered as derived from these basic concepts. The classification of geometric concepts is essential for the formalization of geometric knowledge (see Section 3.3).

3.2 Organization

The organization phase consists in digging out the logical relations among different knowledge segments in standard textbooks in order to provide rules to assist and guide the author to decide the structure of his/her textbook in the process of textbook creation. For each textbook, the author has to find a *linear* order for its knowledge segments (contents) such that their logical relations are satisfied. The segments are then arrayed one by one from the beginning to the end of the textbook for *display* according to this order.

To write a textbook, one has many necessary rules for organizing the textbook data. A human writer may first decide what knowledge to be included in the textbook, then decide the global chapter-section structure to group the knowledge segments, and finally decide the local structure, so that all necessary rules are obeyed. Although one can make his/her own decision as what knowledge to be chosen and how to organize the knowledge, there are some common practice and implicit convention in the scientific community as how mathematical knowledge should be formulated and presented. In order to assist the user to check careless errors in the textbook automatically, the system has to prevent the user from constructing unsound textbooks. For instance, the textbook may be constructed only by using the provided templates and its structure should satisfy some general rules. One general rule for organizing knowledge segments is that the definition segment for every concept in a geometric statement (such as knowledge statement, exercise statement, or example statement) should have been given before this statement. For instance, the definition of altitude should be arranged before the definition of orthocenter, and so should the definition of equilateral triangle before Steiner's theorem. There are many other rules: for example, the fact that the three altitudes of a triangle are concurrent should be given before the definition of orthocenter, and the facts used in the proof of a theorem should appear before this theorem. Of course, the user is permitted to introduce his/her own favorite rules, which have priority over the general rules, for arranging the segments.

Based on these rules, the system can verify the structure of the textbook automatically.

3.3 Representation

We need to provide suitable data representation models for segments and blocks and a standard language for the formalization and representation of knowledge, so that (formalized) knowledge can be translated into natural languages and processed by other software modules.

Knowledge Representation. Geometric statements for concepts, definitions, axioms, theorems, lemmas, corollaries, formulas, properties, etc. are the main contents of geometric knowledge. Predicate logic, a convenient language for the formalization of mathematics, is a natural candidate for the formalization of geometric statements. We use a kind of first-order logic with equality as the representation language. As mentioned before, geometric statements are constructed on the basis of concepts, so formalizing geometric concepts is one of the main tasks.

Taking compound situations of geometric statements into consideration, we assign a relevant *type* to each concept according to their classification discussed in Section 3.1. For instance, the type of orthocenter is Point and the type of altitude is Segment. The concepts of *geometric objects* in plane Euclidean geometry have seven types: Point, Segment, Ray, Line, Angle, Triangle, and Circle. *Geometric quantities* may have four types: Length (e.g., the distance between two points), Degree (e.g., the size of an angle), Number (e.g., the slope of a line), and Area (e.g., the area of a triangle). All these concepts can be considered as *functions* because they are actually nominal and can be used as arguments or terms. The concepts of *object relations* and *quantity relations* have only one type Boolean and can be considered as *atomic formulas* because they are used to declare or decide the truth values of statements.

For each geometric object or quantity, we use its natural name as its *function symbol* and other geometric objects of certain type related to the object or quantity as its function arguments. For example, triangle [A, B, C] represents a triangle with points A, B, C as its vertexes; circle [A, B, C] represents a circle passing through points A, B, C; area [circle [A, B, C]] represents the area of circle [A, B, C]; ratio [length [segment [A, B]], length [segment [B, C]]] represents the ratio of the length of segment [A, B] to the length of segment [B, C]. Here as convention, capital letters are used to represent points, so the function "point" may be omitted by default. Those symbols without function names may be considered as variables.

For each object or quantity relation, we use the short form of its natural name as its *predicate symbol*; its predicate terms are the formalized concepts described by the relation. For example, pon [P, line [B, C]] declares that point P is on line [B, C].

In this way, we can give a formalized expression for each concept:

return Type `Function/Predicate symbol [Type arg1, Type arg2, ...]`

which looks like the definition of a function expression in programming language. For example, "Point intersection [Line a, Line b]" represents the intersection of line a and line b, with return type Point.

We can construct compound statements by composing "function expressions" of this kind, provided that the type constraints are satisfied. For example, "per-Line [intersection [line [A, B], line[C, D]], line [E, F]]" means the perpendicular line from the intersection of line AB and line CD to line EF.

There are many statements (e.g., three altitudes of a triangle) in which we speak about several concepts together. We use a new type Set to indicate this situation with an affiliated piece of information: the number of elements in the set. Therefore, "altitudes [triangle [A, B, C]]" means the three altitudes of triangle ABC and it returns the set and the number (i.e., 3) of segments. Furthermore, "Be[{segment [A, D], segment [B, E], segment [C, F]}, altitudes [triangle [A, B, C]]]" means that AD, BE, CF are the three altitudes of triangle ABC.

In this way, one can formalize every atomic formula in a geometric statement. Formalized geometric statements can be fully constructed by connecting atomic formulas with *connectives* such as ",", "∧" (and), and "⇒" (imply). To make this formal language natural and expressive, the user is permitted to use standard symbols such as "∗" (multiplication), "+" (addition), and "=" (equality) to represent algebraic operations.

Remark: The formalism described above is called a *human-oriented* language because statements in this formalism can be translated without much effort into different natural languages (see Section 4.1) used in traditional textbooks. However, it may be somewhat difficult for other software modules such as theorem provers and diagram drawers to process statements in this human-oriented language because they may not "recognize" so many concepts used therein. To make the conversion easier for communication, it is necessary to provide a *machine-oriented* language as "intermediary" to re-represent geometric statements. Such a language is also in first-order logic and may use only a subset of the concepts (e.g., function and predicate symbols) that are used in the human-oriented language and both of them express the same statements semantically. The machine-oriented representation may be more verbose, but the semantic meanings of the statements may become clearer for other applications. It seems that using the human-oriented language to express the statements which are also needed to be translated into natural languages for display is cumbersome and storing statements in natural languages directly for display is more convenient. However, the purpose of introducing the human-oriented language is to make it possible to automatically "interpret" the meanings of the statements newly created into machine-oriented statements and "translate" them into natural languages when the database is extended (see Section 4.2 for more details).

Data Representation Model. Data representation models refer to various templates that are used to define the contents of classes mentioned in Section 3.1.

Segments and blocks are constructed according to such models. In what follows we provide specifications for some of the models.

Concept {

- **Name (key words)**: function symbol or predicate symbol.
- **Argument/Term list**: (Type arg1, Type arg2, ...) (not necessarily one because, e.g., a circle may be represented by three points or one point and its radius).
- **Argument constraint**: sometimes only the constraints for the types of arguments may be not enough. For instance, when median [X, triangle[A, B, C]] is used to express the median through X, X should be one of the vertexes of triangle ABC.
- **Return type**: type of the defined concept.
- **Pattern**: define patterns for translating the formalized concept into natural languages like English and Chinese.
- **Algebraic expression**: coordinates, equations, or inequations.
- **Hidden information**: when using triangle [A, B, C] in a geometric knowledge statement, we have additional information such as point A, B, C are the vertexes and the opposite side of A is side BC. Information of this kind should be generated automatically during knowledge processing.
- **Nondegeneracy condition**: provide nondegeneracy conditions for the concept (e.g., the vertexes of a triangle are not collinear).

}

Definition Block {

- **Name (key words)**: name of the defined concept.
- **Human-oriented expression**: formalized definition statement in human-oriented language.
- **Machine-oriented expression**: formalized definition statement in machine-oriented language.
- **Consistency condition**: provide geometric constraints for the defined concept. For instance, when the orthocenter of a triangle is used, the fact that the three altitudes of the triangle are concurrent should have been given. This serves to collect relations among different segments of knowledge.
- **Note Segment**: give some background notes or explanations on the definition.
- **Graph Segment**: visualize the defined concept.
- **Example Segment**: give examples illustrating the definition.
- **Exercise Segment**: give exercises related to the definition.

}

Axiom Block: containing Name (key words), Human-oriented expression, Machine-oriented expression, Note Segment, Graph Segment, Example Segment, Exercise Segment.

Theorem/Lemma/Corally/Formula/Property Block: containing Name, Human-oriented expression, Machine-oriented expression, Note Segment, Graph Segment, Example Segment, Exercise Segment, Proof/ Computation Segment.

Note Segment: containing Text, Formalized expression, Link (to literature, graphs, used knowledge in the textbook, etc. to describe the background or explain certain knowledge).

Example Segment: containing Human-oriented expression, Machine-oriented expression, Graph Segment, Proof/Computaion Segment.

Exercise Segment: containing Type of the question (proving, computing, drawing, etc.), Human-oriented expression, Machine-oriented expression, Graph Segment, Solution (Proof/Computation) Segment.

Graph Segment: giving an image or a frame for displaying (dynamic) graphs (where other diagram drawers may be used).

Proof/Computation Segment: giving the process of proving/computing (where other software modules may be used).

Remark: The aim of building a textbook database that collects predefined segments and blocks is to help the user write or create his/her own textbook documents and to keep the collected data reusable. Although the database may provide a lot of formalized knowledge that can be used and processed by other software modules, the data are only formal representations of knowledge statements and do not serve as inference rules as in a deduction system. In other words, the textbook system itself is not a deduction system. Various kinds of operations on the textbook database such as browsing the database, searching for items, modifying the data, and adding new data should be provided. Meanwhile, semantical verification (such as checking for the grammar of formalized expressions, correctness, redundancy, and completeness) during these operations should be performed (see Section 5.1).

4   Textbook Data Processing

4.1   Manipulation

Based on the textbook database, the user can create and edit his/her own textbook documents. The created documents are similar to LaTeX documents in format, but our EGT system will create a dynamic and interactive GUI (instead of static dvi files generated by LaTeX) to display the textbook from the source document. The user may *manipulate* the textbook through the interface visually and *print* out the textbook at any time.

Editing Language. It is expected that the editing language for the textbook document will combine several languages into a new format that presents two aspects of the document — structure and content — and that is usable in other applications.

Structure markups: Declare the construction of segments, blocks, sections, or chapters.

Content definition: Use a data definition language to create the contents of the declared segments or blocks according to the data representation models in Section 3.3.

Query language: As the textbook database collects predefined segments, a query language should be used to fetch the needed segments from the textbook database.

Communicating commands: When some geometric knowledge in the textbook is processed by other software modules, the produced results (such as graphs or proofs of theorems) should be retrieved and integrated into the corresponding blocks. This is handled by using specific commands (such as draw or prove) to communicate with the external software modules.

Interactive Manipulations. The textbook GUI (that displays the processed results in traditional style) is generated by executing relevant query and communicating commands, translating formalized expressions into natural language, and presenting the final data in a configured view. The index of the textbook may be extracted from the document automatically. The following three kinds of operations on the textbook need be considered.

Modifying the contents of the textbook. The contents of the textbook document come mainly from two sources: the textbook database and data newly created by the user. The user may use the data from the database when creating his/her textbook document, but the available data may be insufficient and some of them may be inadequate for his/her purposes. He/she may need to modify the contents of the textbook (such as change the translating style of formalized concepts, modify some texts, and add formalized propositions, examples, and/or exercises) according to his/her interest and need.

Restructuring the textbook. As mentioned before, the textbook data are created by constructing separated segments and blocks. The textbook may be restructured by moving segments or blocks from one place to another.

Setting the style (such as font size, color, and position) of segments.

Translator. Formalized geometric statements of definitions and theorems, etc. may be translated by a translator into natural languages presented in natural style. The translator works at two levels: first translate formalized atomic formulas and then construct geometric statements in a human-readable style by connecting them with connectives and punctuation marks. The atomic formulas are constructed by composing formalized concepts, so the basis is to translate all the concepts. We introduce a *pattern* for each concept and store it in the representation model. For example, the pattern of "perpendicular [arg1, arg2]" is "arg1 is perpendicular to arg2". This is represented by:

$$\frac{\text{perpendicular[arg 1, arg 2]}}{\text{arg 1 is perpendicular to arg 2}} \longrightarrow$$

Given a concrete atomic formula like line [A, B], the pattern-substitution process can be seen as:

$$\frac{\text{line [A, B]}}{\text{line arg 1 arg 2}} \longrightarrow \text{line AB}$$

For compound concepts, translation is performed in a recursive way until the formulas are variables. For example, to shorten the formulas, let

$$\triangle := \text{triangle [A, B, C]},$$
$$\alpha := \text{sides } [\triangle],$$
$$\beta := \text{perLine } [O, \alpha],$$
$$\gamma := \text{foot } [\beta, \alpha].$$

Then translation of the formula Collinear [γ] proceeds as follows:

$$\frac{\text{Collinear } [\gamma]}{\text{arg1 are collinear}} \longrightarrow \gamma \text{ are collinear}$$

$$\frac{\text{foot } [\beta, \alpha]}{\text{the foot/feet of arg1}} \longrightarrow \text{the feet of } \beta$$

$$\frac{\text{perLine } [O, \alpha]}{\text{the perpendiculars from arg1 to arg2}} \longrightarrow \text{the perpendiculars from O to } \alpha$$

$$\frac{\text{sides } [\triangle]}{\text{the three sides of arg1}} \longrightarrow \text{the three sides of } \triangle$$

$$\frac{\text{triangle [A, B, C]}}{\text{triangle arg1arg2arg3}} \longrightarrow \text{triangle ABC}$$

As A, B, and C are variables, we arrive at the complete statement: the feet of the perpendiculars from O to the three sides of triangle ABC are collinear.

For different natural languages (e.g., Chinese, English, and French), different patterns should be provided. Moreover, each chosen pattern should be adjusted according to the type of the arguments. Consider, for example, "foot [arg1, arg2]". If the type of the arguments is Set, then the pattern should be "the feet of arg1"; otherwise, the pattern should be "the foot of arg1".

After all the atomic formulas in a formalized geometric statement have been translated, the whole sentence can be constructed by adding appropriate connectives or punctuation marks. For example, the formalized statement of Simson's theorem is

$$\text{Collinear } [\gamma] \Longleftrightarrow \text{pon } [O, \text{circumcircle } [\triangle]].$$

By translating the involved atomic formulas and inserting appropriate connectives, we can construct the whole statement in English as follows: The feet of the perpendiculars from O to the three sides of triangle ABC are collinear if and only if O is on the circumcircle of triangle ABC. There may be other connectives like "if ... then ..." in a theorem and "... is called ..." in a definition. Connectives are chosen appropriately according to the structures of the atomic formulas.

The use of patterns makes it possible to translate any given formalized geometric statement within the system and to extend the database easily by adding new concepts and relevant patterns. The textbook can even be considered as a "self-translating" system.

4.2   Computation and Deduction

Computation and deduction may be considered as special operations on the knowledge in the textbook. Communication between the textbook and other available geometry software packages such as theorem provers and diagram generators to process the knowledge represented in our formal language is necessary. It is expected that eventually all the lemmas, propositions, and theorems stated in the textbook can be automatically proved, all the geometric quantities introduced in the textbook can be computed, and all the operations defined in the textbook can be performed in real time with simple mouse clicks. The main task is the conversion between the formalized knowledge in our textbook and the input/output formats of the chosen software packages (such as Gool [10]).

Considering that most dynamic geometry systems and reasoning methods use only a subset (basic concepts) of geometric concepts that are used in standard textbooks, we propose a machine-oriented language as "intermediary" (see Fig. 3). Knowledge statements created in the human-oriented language can be automatically interpreted into the machine-oriented language; the latter can then be translated more easily into the formats used in other software modules.

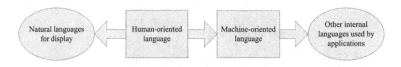

Fig. 3. Translation Steps

Like translation into natural languages by patterns, we use the formal definitions of concepts within the system to eliminate "redundant" concepts in geometric statements through rewriting techniques. This allows human-oriented geometric statements to be "interpreted" into machine-oriented ones. In this sense, the textbook can be considered as a "self-interpreting" system.

For instance, we can "interpret" the hypothesis "Collinear [γ]" of Simson's theorem as follows, where γ is defined as above. First, eliminate the concepts of type Set in the formula. Using "sides [△] := {segment [A, B], segment [B, C], segment[A, C]}", we obtain a formula "Collinear [foot [perLine [O, segment [A, B]], segment

[A, B]], foot [perLine [O, segment [B, C]], segment [B, C]], foot [perLine [O, segment [A, C]], segment [A, C]]]". Next, eliminate the derived concepts "foot" and "perLine" using basic concepts and new variables of points. As an example, we deal with "foot [perLine [O, segment [A, B]], segment [A, B]]". Introduce a new variable point E. Then we have the following sequence of interpretations:

E = foot [perLine [O, segment [A, B]], segment [A, B]]

⇊ the definition of "foot"

E = intersection [perLine [O, segment [A, B]], segment [A, B]]

⇊ the definition of "intersection"

pon [E, perLine [O, segment [A, B]]] ∧ pon [E, segment [A, B]]

⇊ introduce ℓ = perLine [O, segment [A, B]]

pon [E, ℓ] ∧ pon [E, segment [A, B]]

⇊ the definition of "perLine"

pon [E, ℓ] ∧ pon [O, ℓ] ∧ perpendicular [ℓ, segment [A, B]] ∧ pon [E, segment [A, B]]

⇊ the definition of "line"

ℓ = line [E, O] ∧ perpendicular [ℓ, segment [A, B] ∧ pon [E, segment [A, B]]

⇊ eliminate "ℓ"

perpendicular [line [E, O], segment [A, B]] ∧ pon [E, segment [A, B]]

Therefore, we can obtain a set of formulas without "derived concepts", which can be more easily translated into other formats for application.

5   Implementaional Issues and Prototyping

We choose Geometry Revisited [6] as the model of our electronic textbook and Java as the programming language. The object-oriented feature of Java enables the relation between "class" and "object" to be described effectively in the textbook. Java also provides rich primitives for graphical interface implementation. We have implemented a preliminary prototype of the textbook system. The development of the system consists of two main tasks explained below.

5.1   Textbook Database Construction

Data Module. The textbook database collects predefined segments for textbook construction. These segments are considered and constructed as objects of the corresponding classes, such as Definition class, Axiom class, Lemma class, Theorem class, Corollary class, Property class, Formula class, Note class, Example class, Exercise class, Graph class, and Proof/Calculation class. The definition of these classes, i.e., the templates for constructing segments, need be provided at first. The contents of *Definition Class* and *Theorem Class* may be specified as

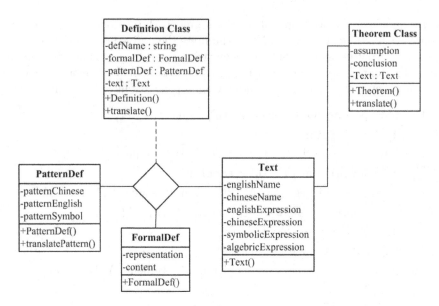

Fig. 4. Templates of Definition and Theorem Segment

in Fig. 4. We have constructed 48 definition segments and 17 theorem segments inside the system according to these "templates."

Managing these segments for easy browsing, searching, retrieving, and processing is another issue. In the current prototype, the predefined segments are organized and stored under Java Set Framework. Each segment is assigned a unique key: the defined concept name of a definition or the well-known name of a theorem is used as the key of reference to the corresponding segment. For such theorems that do not have any well-known names, unique keys are also created and assigned for reference inside the system.

Operation Module. Based on the keys assigned to segments, searching a target segment such as the definition of orthocenter or the theorem of Simson by its well-known name can be easily implemented. However, in most situations, the name of the target segment is not clear, and thus searching at the semantic level is important and necessary. For example, we know the statement of a definition but not its name (as some theorems even have no well-known names) and we may wish to find theorems whose hypotheses or conclusions are the same semantically, but stated in different ways. With this kind of searching service, the user may obtain real-time guide to extract the needed segments from the database or find the needed data while editing the textbook documents. The questions of how to define this kind of query language and how to execute the search remain for further investigation.

Moreover, interactive operations on the database such as modifying the contents of segments and constructing and adding new segments into the database (with dialogues) need be provided for the user to easily maintain the database.

These operations have not yet been implemented, but are on the top of our stack of programming tasks.

Verification Module. Since the segments stored in the database contain "geometric meanings" semantically, verification should be performed when modifications on the database are operated. For example, the following properties should be checked.

Grammar: when geometric statements are formalized, the construction of formalized expressions need obey some rules (e.g., the type constraints). The constructed expressions can be processed if and only if they are grammatically correct.

Correctness: check the correctness of geometric propositions and the possibility of declaring a geometric statement. This may work by employing other software modules such as theorem provers or geometric constraint verifiers.

Redundancy: check the redundancy of the database when new data are added. For example, knowledge segments already existing in the database should not be added again.

Completeness: check whether all the segments that *derive* the newly added segments already exist in the database. This ensures the completeness of the database.

A proposed working frame of these three modules is depicted in Fig. 5. When a modification on the Data Module is generated from the Operation Module, the Verification Module will check whether it is permitted. If the modification is verified successfully, it will be performed. Otherwise, Warning message will be produced, explaining the cause to the user.

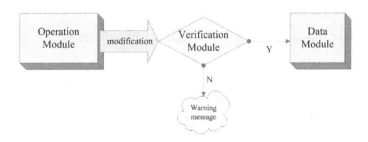

Fig. 5. Textbook Database Frame

5.2   Document Creation and Textbook Manipulation

To create the textbook by using, organizing, and processing the data from the database, we need to implement various functions. Such functions may be grouped into several modules.

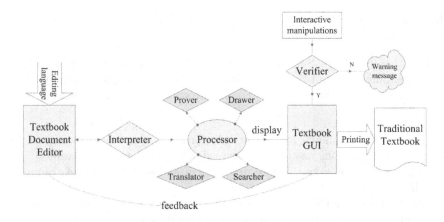

Fig. 6. Document Creation and Textbook Manipulation

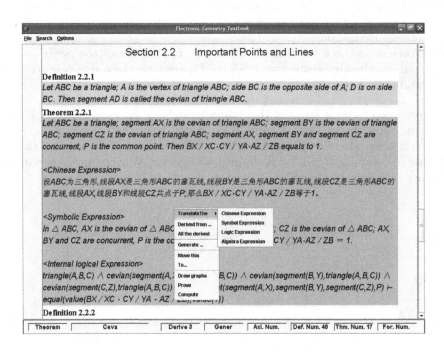

Fig. 7. Textbook GUI

The frame of these modules that we propose is shown in Fig. 6. The user creates his/her textbook document in Textbook Document Editor environment by using the Editing language discussed in Section 4.1. The document will be interpreted by the Interpreter. If no error occurs, the interpreted information will be transported into the Processor, which will perform operations such as translating the formalized statements into natural languages, querying the needed segments

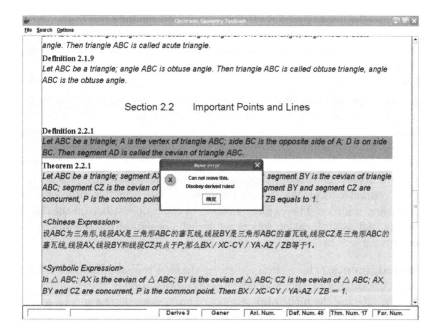

Fig. 8. Try to Move the Ceva Theorem Before the Definition of Cevin

for use, and communicating with other software modules to prove a theorem or generate a diagram. The Textbook GUI will then be created to present or display all the contents produced by the Processor. Interactive manipulations such as modifying the contents of segments, proving a theorem, generating a diagram, and reorganizing the structure of the textbook will be checked by the Verifier. If the verification is passed successfully, the manipulation will be performed on the GUI and feedback about the modified information will be transported to the textbook document. Otherwise, a Warning message will be generated and the manipulation will be denied. Furthermore, the textbook may be printed out as a Traditional Textbook at any time.

We have implemented a translator for translating formalized statements into English and Chinese, a GUI for displaying constructed segments in natural languages (see Fig. 7), and some functions for manipulating and restructuring knowledge segments. The operability of restructuring can be verified according to the rules proposed in Section 3.2 (see Fig. 8: if some rules are disobeyed, then the incompatible segments will be highlighted; if all the rules are obeyed, then the label for each segment will be adjusted automatically).

6   Conclusion and Future Work

In this paper, we have presented the design of a dynamic interactive (geometry) software system that integrates available recourses into a uniform environment — *textbook* — and that realizes textbook data standardization, share and reuse.

The user can edit his/her own textbook documents, making use of the textbook database, and manipulate them visually and interactively. The built-in geometric knowledge can be processed by different software modules and be presented in natural style for viewing and printing. The system may be used for education, research, and technical document writing and processing.

The design and implementation of algorithms and functions for the tasks discussed above and the construction of the GUI that integrates different interactive functions require an enormous amount of work and effort. Currently we are still at the early stage of building the system. Based on the implemented prototype, we will provide the definitions of other classes, construct more segments into the textbook database, and make experiments to find efficient techniques for querying at the semantic level. We will also implement functions for interaction with other software modules. The implementation of other operations on the database and GUI as well as verification functions and the design of the document editor and interpreter will be carried out at a later stage.

Acknowledgments. The authors wish to thank Professor Bruno Buchberger and other organizers of the Special Semester on Gröbner Bases and Related Methods for offering them the opportunity to participate in the special semester. This work has benefited considerably from the authors' communications with Bruno and other members of the Theorema group during their stay in Linz. The authors also wish to thank the referees for their insightful comments and suggestions which have helped bring this paper to the present form. This work is financially supported by the National Key Basic Research Projects 2004CB318000 and 2005CB321902 of China.

References

1. ActiveMath Home, http://www.activemath.org/
2. Allen, S., Bickford, M., Constable, R., Eaton, R., Kreitz, C., Lorigo, L.: FDL: A Prototype Formal Digital Library. Cornell University, USA (2002), Available at http://www.nuprl.org/documents/FDL/02cucs-fdl.pdf
3. Asperti, A., Padovani, L., Coen, C.S., Schena, I.: HELM and the Semantic Math-Web. In: Boulton, R.J., Jackson, P.B. (eds.) TPHOLs 2001. LNCS, vol. 2152, pp. 58–74. Springer, Heidelberg (2001)
4. Caprotti, O., Carlisle, D.: OpenMath and MathML: Semantic Mark Up for Mathematics. ACM, New York (1999), ACM Crossroads, http://www.acm.org/crossroads/xrds6-2/openmath.html
5. Cinderella Home, http://www.cinderella.de/
6. Coxeter, H.S.M., Greitzer, S.L.: Geometry Revisited. The Mathematical Association of America, Washington, DC (1967)
7. Janičić, P.: GCLC — A Tool for Constructive Euclidean Geometry and More than That. In: Iglesias, A., Takayama, N. (eds.) ICMS 2006. LNCS, vol. 4151, pp. 58–73. Springer, Heidelberg (2006)
8. Kohlhase, M.: OMDoc: An Infrastructure for OpenMath Content Dictionary Information. ACM SIGSAM Bulletin 34(2), 43–48 (2000)

9. Kohlhase, M., Franke, A.: MBase: Representing Knowledge and Context for the Integration of Mathematical Software Systems. J. Symb. Comput. 23(4), 365–402 (2001)

10. Liang, T., Wang, D.: Towards a Geometric-Object-Oriented Language. In: Hong, H., Wang, D. (eds.) ADG 2004. LNCS (LNAI), vol. 3763, pp. 130–155. Springer, Heidelberg (2006)

11. Lozier, D.W.: NIST Digital Library of Mathematical Functions. Ann. Math. Artif. Intell. 38(1-3), 105–119 (2003)

12. Piroi, F., Buchberger, B.: An Environment for Building Mathematical Knowledge Libraries. In: Windsteiger, W., Benzmueller, C. (eds.) Proceedings of the Workshop on Computer-Supported Mathematical Theory Development, Cork, Ireland, pp. 19–29 (2004)

13. Quaresma, P., Janičić, P.: GeoThms — Geometry Framework. Technical Report 2006/002, Centre for Informatics and Systems, University of Coimbra (2006)

14. Trybulec, A., et al.: The Mizar System. Available and developed at the University of Warsaw, Poland, at `http://mizar.uwb.edu.pl/system/`

15. Wang, D.: GEOTHER 1.1: Handling and Proving Geometric Theorems Automatically. In: Winkler, F. (ed.) ADG 2002. LNCS (LNAI), vol. 2930, pp. 194–215. Springer, Heidelberg (2004)

16. W3C Math Home: What is MathML?, `http://www.w3.org/Math/`

17. Zeilberger, D.: Plane Geometry: An Elementary Textbook by Shalosh B. Ekhad, XIV (circa 2050), downloaded from the future by Doron Zeilberger. Available from: `http://www.math.rutgers.edu/~zeilberg/GT.html`

Equidecomposable Quadratic Regions*

Thomas C. Hales

Math Department, University of Pittsburgh, Pittsburgh, PA 15217
hales@pitt.edu

Abstract. This article describes an algorithm that decides whether a region in three dimensions, described by quadratic constraints, is equidecomposable with a collection of primitive regions. When a decomposition exists, the algorithm finds the volume of the given region. Applications to the 'Flyspeck' project are discussed.

1   Introduction

From ancient times, a common approach to computing the volume of a region A is to dissect A into finitely many pieces and then reassemble those pieces into finitely many new regions whose volumes have been previously determined.

This is the motivating idea behind Hilbert's third problem on polyhedra. He asked if any two polyhedra of the same volume are equidecomposable. That is, can one polyhedron be cut into finitely many polyhedral pieces, which can be reassembled into the other polyhedron. M. Dehn answered this question negatively in 1902. For example, a regular tetrahedron is not equidecomposable with a cube of the same volume. The proof of this result is that the cube and the tetrahedron have different Dehn invariants, but all equidecomposable polyhedra have the same Dehn invariant.

In two dimensions, the corresponding result is true: any two polygons of the same area are equidecomposable. That is, if they have the same area, the first polygon can be cut into finitely many triangles in such a way that they can be reassembled into the second polygon.

The mathematical literature on equidecomposability has various extensions. The Banach-Tarski paradox is one of the best known of these results: it is possible to cut a ball into finitely many (non-measurable) pieces and reassemble them into a two balls of the same radius as the first. Another is the classical squaring-the-circle problem: M. Laczkovich proved that it is possible to cut a square into finitely many pieces that can be reassembled into a disk of the same area [6].

In this article we look at regions that are defined by a restricted class of quadratic constraints. Each face of the boundary is required to be planar, spherical, or right conical. We call these quadratic regions. Among the quadratic regions are a special subset that we call primitive. All the primitive regions are familiar shapes. The question we ask is when a quadratic region is equidecomposable with a finite disjoint collection of primitive quadratic regions. When such a decomposition can be produced, we obtain a formula for the volume of the quadratic region in terms of primitives.

---

* This research has been supported by NSF grant 0503447.

F. Botana and T. Recio (Eds.): ADG 2006, LNAI 4869, pp. 24–38, 2007.
© Springer-Verlag Berlin Heidelberg 2007

This article describes an algorithm to decide whether a quadratic region admits such an equidecomposition and to produce the decomposition when it exists. Our analysis of the algorithm is reminiscent of the Dehn invariant, which is an invariant attached to the edges of a polyhedron. Similarly, most of our analysis is focused on the curves formed by the intersection of two quadric surfaces.

The study of intersections of conics is classical. In Descriptive Geometry, this was one of the topics "which possess industrial utility and which develop the qualities of mind so essential in a draftsman" [5]. More recently, the intersection of quadric surfaces has become relevant for Computer Aided Geometric Design. For a guide to the recent literature on this subject, see the survey in the introduction to [2].

This algorithm has applications to the proof of the Kepler conjecture, and the ongoing project to formalize that proof. All of the volume calculations that arise in the proof of the Kepler conjecture are volumes of quadratic regions. In every case, these regions are equidecomposable with a finite disjoint collection of primitive regions. Thus, all of the volume calculations of that proof can be handled by an automated procedure.

2 Petal Figures

To motivate what is to come, we begin with the simple case of two dimension petal figures. To limit ourselves to the area of bounded regions in the plane, we introduce a large square $\Omega = (-L, L)^2$, which will function as our "universe." When we speak of the complement of an open set $A \subset \Omega$, we mean the set $\Omega \setminus A$.

Definition 1. *A petal is a convex region in Ω, whose boundary is formed by a finite set of line segments and arcs of circles. A petal figure is a finite union of petals and complements of petals.*

The boundary of a petal has zero area, so it does not matter for our purposes whether it is included in the petal or not. We will disregard sets of measure zero.

We define the primitive petals to be two different types of sets, which we will call *caps* and *triangles* respectively:

1. Cap: the part interior to a circle, bounded by the circle and a line intersecting the circle.
2. Triangle (including the interior).

Let $\chi(A)$ be the characteristic function of a set A. All characteristic functions will be considered modulo the subspace of functions generated by characteristic functions of sets of measure zero.

Problem 1. Given a petal figure A, express $\chi(A)$ as an integral linear combination of characteristic functions of primitive petals.

The solution to this problem is a simple matter. Draw the secant to each circular arc forming part of the boundary of A, and take the corresponding caps C_i. Let $\sigma_i = 1$, if locally along the circular arc, A and C_i lie on the same side of the arc. Set $\sigma_i = -1$ otherwise. Working at the level of characteristic functions, we see that $\chi(A) - \sum_i \sigma_i \chi(C_i)$

is a linear combination of characteristic functions of polygons. By triangulating each polygon, we obtain triangles T_j. Then we obtain

$$\chi(A) = \sum_i \sigma_i \chi(C_i) + \sum_j \chi(T_j).$$

2.1 Petal Figures on a Sphere

We can repeat this construction, using a sphere instead of the plane. We take Ω to be the unit sphere, and we define a spherical petal figure to be a finite boolean combination (intersections, unions, and complements) of regions bounded by circles on the unit sphere.

We take the primitives to be the interior of geodesic (great circle) triangles and the two-sided regions bounded by an arc of a circle and an arc of a great circle.

Following the same procedure, we can rewrite the characteristic function of a spherical petal function as a linear combination of characteristic functions of primitives.

3 Primitive Volumes

Definition 2. *The open ball $B(x, r)$ with center x and radius r is the set*

$$\{y \in \mathbb{R}^3 \mid |x - y| < r.\}$$

Definition 3. *The solid lune $L(x, v, w)$ is specified by a point $x \in \mathbb{R}^3$, and two unit vectors v and w. It consists of points y such that*

$$(y - x) \cdot w > 0 \ \wedge \ (y - x) \cdot v > 0.$$

This is the intersection of two half-spaces.

Definition 4. *The solid triangle $ST(x, r, v_1, v_2, v_3)$ is specified by a point $x \in \mathbb{R}^3$, a radius $r \geq 0$, and unit vectors v_1, v_2, v_3. It is the intersection of two lunes with a ball:*

$$ST(x, r, v_1, v_2, v_3) = L(x, v_1, v_2) \cap L(x, v_2, v_3) \cap B(x, r).$$

The vector v_2 is repeated, so it has three bounding planar faces, rather than four.

Definition 5. *The solid cap $SC(x, v, r, c)$ is specified by an apex $x \in \mathbb{R}^3$, a radius $r \geq 0$, a non-zero vector v giving direction, and constant c. The solid cap is the intersection of the ball $B(x, r)$ with a half-space:*

$$\{y \in B(x, r) \mid (y - x) \cdot v > c\}.$$

Definition 6. *The frustum $FR(x, v, h', h, a)$ is specified by an apex $x \in \mathbb{R}^3$, heights $0 \leq h' \leq h$, a unit vector v giving its direction, and $a \in [0, 1]$. The set FR is given as*

$$\{y \mid (y - x) \cdot v > a|y - x| \ \wedge \ h' < (y - x) \cdot v < h\}.$$

By squaring the first inequality, we get the equation of a frustum as a quadratic constraint:

$$((y - x) \cdot v)^2 > a^2((y - x) \cdot (y - x)).$$

Definition 7. *A tetrahedron* $S(v_1, \ldots, v_4, c_1, \ldots, c_4)$ *is the intersection of four open half-spaces (when the intersection is nonempty).*

$$y \cdot v_i < c_i, \quad i = 1, 2, 3, 4.$$

These sets have all been defined by linear and quadratic constraints.

Definition 8. *A primitive region is any of the following.*

1. *A solid triangle* ST.
2. *A tetrahedron* S.
3. *A wedge of a frustum; that is, the intersection of a lune with a frustum:*

$$FR(x, u, h', h, a) \cap L(x, v, w),$$

   where $h' < h$ *and* $u \cdot v = u \cdot w = 0$.
4. *A wedge of a solid cap; that is, the intersection of a lune with a solid cap:*

$$SC(x, r, u, h) \cap L(x, v, w), \quad \text{where } u \cdot v = u \cdot w = 0.$$

The solid triangle has one face that is spherical. The curvilinear edges of that face are arcs of great circles. A wedge of a solid cap also has one face that is spherical. The curvilinear edges of that face are circles. Two edges are great circles and the third edge is not necessarily a great circle. A wedge of a frustum has one face that is conical. Two of the edges of that face are line segments. The other edges are arcs of a circle (a conic section).

The volumes of these primitive regions are well-known. We will not repeat them here. These are elementary integrals. The volume formula for a solid triangle goes back over 400 years to T. Harriot.

4 Statement of Results

The main problem in rough terms is to determine whether a given quadratic region is equidecomposable with a finite disjoint sum of the primitive regions. This section gives a precise formulation of the problem.

If the coefficients of the linear and quadratic constraints used to define a primitive region are algebraic numbers, then we say that the region is definable. We restrict ourselves to definable sets, to avoid computations with arbitrary real numbers.

Let $\mathcal{A}$ be the collection of all sets A obtained by finite boolean combinations (intersection, union, and difference) of definable primitive regions. The operations of intersection, finite union, and difference carry bounded sets to bounded sets. We call the sets in $\mathcal{A}$ quadratic regions. Let $\mathcal{F}$ be the vector space of all rational linear combinations of characteristic functions of sets $A \in \mathcal{A}$ (considered modulo rational linear combinations of characteristic functions of sets of measure zero).

We define the equivalence relation $f_1 \sim f_2$ of $\mathcal{F}$ generated by $\chi(A) \sim \chi(B)$, if there is a rigid motion of $\mathbb{R}^3$ carrying A to B, with $A, B \in \mathcal{A}$. Let $\mathcal{F}_0$ be the vector space of all rational linear combinations of characteristic functions of definable primitive regions.

Problem 2.   – Given $f \in \mathcal{F}$ (explicitly presented), determine whether it is equivalent
    to a function in $\mathcal{F}_0$.
  – If $f \in \mathcal{F}$ is equivalent to a function in $\mathcal{F}_0$, find a representative of f in $\mathcal{F}_0$.

Equivalent functions have the same (Lebesgue) integral. Thus, an affirmative answer
to the problem for a function f leads to its integral in terms of volumes of primitive
regions.

Theorem 1. *There is an algorithm that answers Problem 2.*

In particular, the algorithm finds a way, if it exists, to move non-primitive pieces by rigid
motions so that they can be assembled into primitives. For instance, start with a single
primitive region, cut it into two regions that separately cannot be expressed in terms
of primitives, then move the pieces apart by rigid motions. The resulting characteristic
function is not in $\mathcal{F}_0$, but it is equivalent to a function in $\mathcal{F}_0$. The algorithm finds the
rigid motion that moves these pieces back together.

The method we use is similar to a familiar method of solving a two-dimensional
jigsaw puzzle, by picking up a single piece at a time, and trying to match its edge with
every other piece in turn, until a match is found. (If several pieces have an identical
edge, extra bookkeeping is involved).

I have made no attempt to find the most efficient possible algorithm. It is quite obvious that significant improvements are possible over what I have presented.

There are various more general problems along these lines that can be posed. There
is no reason to restrict the primitive regions to the particular regions that are used here.
The same problem can be posed with more general collections of primitive regions. I
have not investigated these more general problems. It would be interesting to do so.

5   The Algorithm in Overview

Each $A \in \mathcal{A}$ is bounded by finitely many surfaces. Each surface is planar, spherical,
or conical. Two surfaces intersect along a segment of a curve. Since the surfaces are
quadrics, each curve has degree at most four. If the degree of the curve is one, it is a
line. If the degree of the curve is two, then it is a conic section. In particular it is a
planar curve. When the intersection of a cone and a sphere is reducible, the irreducible
components are planar. When the intersection of two cones is reducible, the intersection
consists either of planar curves or of a line and an irreducible nonplanar cubic. The
algorithm proceeds by making a careful analysis of these curves and the local geometry
near the curve.

The idea is to proceed in stages, starting with the most complicated types of curves,
and using the equivalence relation to rewrite the function f in equivalent form with
curves of lesser complexity. As we progress, the curves appearing in the representation
of the function become progressively simpler, until finally, the curves are all lines and
circles. At this point, we are able to recognize the primitive regions. The algorithm
reports that f is not equivalent to a function in $\mathcal{F}_0$ if at any stage, it is impossible to
eliminate the curves of a given type. The only way that the algorithm fails to find a
representative of f in $\mathcal{F}_0$ is for no such representative to exist.

For each $f \in \mathcal{F}$, we work with a representation of f as a linear combination $f = \sum a_i \chi(A_i)$ of quadratic regions. We will call the elements A_i the constituents of f. As the algorithm progresses we may change the representation of f and hence its constituents.

As the algorithm progresses we will subtract off various known quantities from the function f, until finally it becomes the zero function. Thus, the function f does not denote a fixed element of $\mathcal{F}$, but rather a dynamic quantity that depends on the stage of the algorithm.

We will repeatedly use the identity:

$$\chi(A) = \chi(B) - \chi(B \setminus A) + \chi(A \setminus B).$$

The set A will be the region we are trying to decompose. The set B will be a set of known volume that matches A along a surface and a curvilinear edge of that surface. By matching B along the 'most complex feature' of the region A, the remainder $\chi(A \setminus B) - \chi(B \setminus A)$ will have lower complexity than A, and this will ensure that the procedure eventually terminates.

Although the expressions $\chi(A) - \chi(B)$ and $\chi(A \setminus B) - \chi(B \setminus A)$ are equal as functions, the data that represent them differ, and it matters in the algorithm which expression we use. In fact, it is best to retain the expression $\chi(A) - \chi(B)$ rather than rewrite it in terms of $A \setminus B$ and $B \setminus A$. Rewriting it would require an analysis of the intersection of the boundaries of A and B, and this would introduce further complications.

6 Capabilities

The algorithm will require us to construct regions with certain properties, and it will require us to decide the truth value of various statements. This section collects some of these auxiliary results upon which the algorithm rests.

6.1 Elementary Theory of the Reals

In the course of the algorithm, there are various statements in the elementary theory of the reals whose truth value we need to determine. Since the elementary theory of the reals is decidable, we can make unrestricted use of such in our algorithm. For an overview of methods of quantifier elimination with applications to geometry, see [1] and [7]. For example, we may have a point $x \in \mathbb{R}^3$ and a set $A \in \mathcal{A}$ and ask whether every sufficiently small neighborhood of x is contained in A. This statement can be expressed as a sentence in the first order theory of the reals. In our description of the algorithm, we will make use of such statements, sometimes without explicitly mentioning that they are elementary. When we use the term 'elementary' in this paper, it is in this technical sense.

6.2 Jumps Across Surfaces

Let C be an irreducible component of the intersection of two surfaces:

$$C \subset \{x \mid f_1(x) = f_2(x) = 0\},$$

defined by polynomials f_1 and f_2, where the zero sets Z_i of f_i are planes, spheres, or cones.

Let x be a nonsingular point on C that does not lie on any other irreducible component of the zero set of (f_1, f_2). If the two surfaces are smooth at x and if their tangent planes to the surfaces at x are not equal, then by the implicit function theorem, in sufficiently small balls around x, the topology of the triple (x, C, Z_i) is that of a point, diameter, and plane through the diameter.

If C is nonplanar, these conditions are satisfied for all but finitely many points of C. In fact, C meets another component of $Z_1 \cap Z_2$ with at most finite multiplicity. The number of singular points on C is finite. The set of points x for which the tangent planes are equal is clearly Zariski closed. Hence if the tangent planes are equal at infinitely many x, they are equal for all $x \in C$. Now if the two surfaces are cones, the tangent planes to the curve at three different points, none on the same ray through the apex, determine the apex, and hence determine the cone. Thus, the intersection of two distinct cones cannot give the same collection of tangent planes along the curve. If the two surfaces are a cone and a sphere, the segments from the apex to the curve are tangents to the sphere. Hence they have the same length. This forces the curve of intersection to be a circle, contrary to the assumption that C is nonplanar. So indeed these conditions are satisfied for all but finitely many points of C.

In particular, locally at x, C separates Z_1 into two connected components

$$E(f_1, \pm f_2) = \{x \mid f_1(x) = 0, \ \pm f_2(x) > 0\}.$$

Let A be a quadratic region. For every f_1, f_2, consider the elementary statement:

$P(f_1, f_2, C, x, A)$: *For every sufficiently small open ball $B(x, \epsilon)$ around x and for every*

$$y \in B(x, \epsilon) \cap E(f_1, f_2),$$

there is a nonempty open ball $B(y, \delta) \subset B(x, \epsilon)$ such that

$$B(y, \delta) \cap \{z \mid f_1(z) > 0\} \subset A.$$

Let $\psi(f_1, f_2, C, x, A) = 1$ if $P(f_1, f_2, C, x, A)$ is true and 0 otherwise. If $f = \sum_i a_i \chi(A_i)$, we have a corresponding value

$$J(f, f_1, f_2, C, x) = \sum_i a_i(\psi(f_1, f_2, C, x, A_i) - \psi(-f_1, f_2, C, x, A_i)).$$

This function measures the jump in the value of the function f across the zero set of f_1, on the side $E(f_1, f_2)$. It is elementary to compute it for a given $x \in C$.

If every intersection of $B(x, \epsilon) \cap \{y \mid f_1(y) = 0\}$ with a surface $f_2 = 0$ defining A is contained in C, then except for finitely many $x \in C$, the element y in the statement $P(f_1, f_2, C, x, A)$ does not lie on any surfaces defining A except $f_1 = 0$. (We will only use the jump function when this condition is met.) Thus, the set $B(y, \delta) \cap \{z \mid f_1(z) > 0\}$ for δ sufficiently small is either disjoint from A or is entirely contained in A.

Under these same conditions, the jump $J(f, f_1, f_2, C, x)$ is a locally constant function of x on $C \setminus X$, where X is a computable finite set of exceptional points. The set $C \setminus X$ is a finite number of topological intervals. Thus, $J(f, f_1, f_2, C, x)$ can be considered a function on a finite domain, computed by picking a test point x on each interval. (We have more to say in Section 6.3 about how to specify the topological intervals).

We say that C is a *boundary curve* of $f \in \mathcal{F}$ if there is a jump

$$J(f, f_1, f_2, C, x) \neq J(f, f_1, -f_2, C, x),$$

for all x in some interval of C for some (f_1, f_2). We also say that C is a boundary curve of $A \in \mathcal{A}$, if it is a boundary curve of $\chi(A) \in \mathcal{F}$. We say that $f_1 = 0$ is a *boundary surface* of $A \in \mathcal{A}$ if

$$J(\chi(A), f_1, \pm f_2, C, x) \neq 0.$$

for some choice of sign $\pm$ (again for x in some interval of C), and some f_2 such that C is an irreducible component of the zero set of (f_1, f_2). We say that the jump through $f_1 = 0$ is coherent across C (at x) if

$$J(f, f_1, f_2, C, x) = J(f, f_1, -f_2, C, x),$$

for every f_2 such that C is an irreducible component of the zero set of (f_1, f_2).

6.3   Intervals on a Curve

There are several arguments that make use of a simple arc of an algebraic curve on a quadric surface. Since we do not have parameterizations of the curves, we use the surface to simplify matters.

Suppose that C is a nonplanar curve on a cone S, given as an irreducible component of the intersection of two cones. The lines through the apex and the circles form isothermal coordinates on S that can be used to partition the cone into small 'rectangles.'

We claim that the rectangles can be chosen so that the intersection of C with each rectangle is a simple arc (except at singular points of C and the apex of the cone). Simply pick lines on the rectangular grid that include every tangent line to C that lies on the cone, and circles that include every circle on S tangent to C. On each such rectangle, each branch of C is a continuous function of the isothermal coordinates, and is therefore homeomorphic to an interval. By restricting the size of the rectangles further, we may assume that there is a single branch of C in each rectangle. (Checking that C is a univalent function on a rectangle is an elementary test.)

A similar procedure works for a curve C on a sphere S. We may pick two pairs of antipodal points (that avoid C), and then take a collection of great circles that pass through one pair or the other. This gives a system of 'rectangles' that can be used in a similar manner to break C into a finite set of intervals. We will eventually relate these rectangles to primitive regions. With that in mind, we draw a diagonal to each rectangle, breaking it into two spherical triangles.

The possible singularities of C are mild in the case of the intersection of a cone with another cone or sphere. It can be verified that if there are two branches of the curve C meeting at a singular point x on the curve, then it is always possible to separate the branches in a neighborhood of x by a plane passing through x. This is true even if x is an apex of a cone. We can choose the rectangles so that the singular points of C always lie on an edge of a rectangle. By cutting the rectangles by finitely many planes, we arrive at a situation where we separate all the branches of the curve C. We refer to these rectangles, possibly cut into smaller pieces by branch-separating planes as *charts* for C

(with respect to a sphere or cone S). When the surface is a sphere, by triangulating if necessary, we may assume that each chart is a spherical triangle.

We can choose the charts sufficiently small, so that on the intervals I they define, the jump function J is constant on each interval. We write $J(f, f_1, f_2, C, I)$ for the common value of $J(f, f_1, f_2, C, x)$, for $x \in I \subset C$.

6.4 Automorphisms of Curves

A bijective rigid motion $T : A \to B$ between subsets $A, B \subset \mathbb{R}^3$ is a *congruence*. A congruence of a set A with itself is an *automorphism* of that set. A nonplanar irreducible algebraic curve in $\mathbb{R}^3$ has only finitely many automorphisms.

Given an irreducible nonplanar component C of the intersection of a cone with a sphere or another cone, we can determine the finite group of rigid motions that map the curve to itself. This is elementary. A rigid motion T is determined by the finitely many coordinates of its underlying affine transformation

$$T(x) = Ax + b.$$

The statement 'T *is automorphism of C*' can be expressed in the elementary language of the real numbers, with free parameters A and b. By eliminating quantifiers from that statement, we obtain explicit equations for the possible automorphisms T.

For each automorphism T, we can solve $T(x) = x$ for the set of fixed points, and then intersect the set of fixed points with C to determine the finite set of fixed points of T on C. By passing to smaller charts in Section 6.3, we can assume that each point of C that is fixed by some automorphism is an endpoint of an interval, so that no fixed points appear on an open-ended interval.

We may repeatedly bisect the charts (and the intervals of the curve C they contain) until each (open-ended) interval I is so small that the only automorphism T such that $T(I)$ meets I is the identity automorphism.

6.5 Automorphisms and Boundary Edges

If $f = \sum a_i \chi(A_i) \in \mathcal{F}$, we can form a list of all of its boundary curves $C_1, \ldots, C_k$. By applying quantifier elimination on each of the statements 'T *is a rigid motion carrying C_i bijectively to C_j*', we can explicitly determine all of the finitely many congruences between the boundary curves of f. We can subdivide each boundary curve C_i into small open-ended intervals I_{ir} (that cover C_i except for the finitely many endpoints), as explained in Section 6.4, so that $T(I_{ir})$ does not meet I_{ir}, whenever T is a non-trivial automorphism of C_i.

Fix one boundary curve $C = C_0$, and consider all congruences from the other curves C_i to C (including C_0 itself). Mark all endpoints of the intervals $T(I_{ir}) \subset C$, as we run over all congruences T for all i. Use these finitely many points so obtained, to further refine the intervals I_{0r} into smaller intervals $I'_{0r'}$. If some $I'_{0r'}$ meets some $T(I_{ir})$, then $I'_{0r'}$ is contained in $T(I_{ir})$.

By construction, if $T' : C \to C_i$ is a congruence, then $T'(I'_{0r'})$ is contained in a single interval I_{ir} of C_i. The intervals $I'_{ir'} = T'(I'_{0r'})$ then give a refinement of the intervals I_{ir} that is independent of the choice of $T' : C \rightarrow C_i$. In this way, we obtain a collection of intervals $I'_{ir'}$ on each curve C_i (that cover except for finitely many endpoints) such that every congruence carries intervals bijectively to intervals.

6.6 The Edge Coherence Condition

Let C be an irreducible curve that has only a finite number of automorphisms.

Let $f \in \mathcal{F}$. Let $C_1, \ldots, C_k$ be the boundary curves of f that are congruent to C. Let $\mathcal{T}$ be the set of all congruences $T : C_i \rightarrow C$, as i ranges from 1 to k.

The group of motions of $\mathbb{R}^3$ acts on $\mathcal{F}$ by

$$(T_* f)(x) = f(T^{-1} x).$$

With this action, $T_* \chi(A) = \chi(TA)$. We define $D_C(f) = \sum_{T \in \mathcal{T}} T_* f \in \mathcal{F}$. This is well-defined in the sense that it does not depend on the expression for f as a sum of characteristic functions. Note that D_C depends on f through the list $C_1, \ldots, C_k$ of boundary curves, so it is not a linear operator on $\mathcal{F}$.

If $f = \sum_i a_i \chi(A_i)$. Then

$$D_C(f) = \sum_{T \in \mathcal{T}} \sum_i a_i \chi(TA_i).$$

Partition each C_i into intervals I'_{ir} as described in Section 6.5. Define intervals I_r on C by $I_r = T(I'_{ir})$ for any congruence $T : C_i \rightarrow C$.

We have the following necessary condition for equidecomposability. If the following condition is not met, the function f is not equivalent to a function in $\mathcal{F}_0$, and the algorithm terminates.

(Edge Coherence Condition) For every nonplanar irreducible curve C, every interval I_r of C as constructed above, and every nonzero quadratic function f_1 whose zero set contains C, the function $D_C(f)$ has jumps through $f_1 = 0$ that are coherent across C along I_r.

This is stated as a condition on infinitely many curves and quadratic functions. However, it reduces to a finite calculation. We can restrict to curves C that are congruent to an edge curve of f, because otherwise the condition always holds. We can restrict further, to curves C that equal one of the edge curves $C_1, \ldots, C_k$. We can restrict the condition further to quadratic functions f_1 that have the form $T_* f'_1$, where f'_1 defines a boundary surface of some constituent A_i of f, the zero set of f'_1 contains some C_j, and $T : C_j \rightarrow C$ is a congruence. This reduces the edge coherence condition to a finite condition.

We will omit the proof that any function equivalent to a function in $\mathcal{F}_0$ satisfies the edge coherence condition. There are two main ingredients to the proof. The first ingredient is to check that the edge coherence condition holds with respect to $D_C(f)$ if and only if it holds for a function $D'_C(f)$, where D'_C is defined exactly as D_C, but with respect to a larger set of curves $\{C_1, \ldots, C_k, C_{k+1}, \ldots, C_\ell\}$, where the additional curves are congruent to C, but otherwise arbitrary. The second ingredient is to check that if $A \in \mathcal{A}$ and T' is any rigid motion, the function $f' = f + a(\chi(A) - \chi(T'A))$

satisfies the edge coherence condition if and only if f does. The proof of this second ingredient is based on the first ingredient.

We note that the edge coherence condition is a Dehn invariant type condition. With the original Dehn invariant, all the edges are line segments, which are all locally congruent. This helps to explains why the Dehn invariant, which is a single invariant for polyhedra, must be expanded to a collection of conditions for general quadratic regions. We have a separate condition for each congruence type of curve.

7 Remove Nonplanar Intersections

We are ready to give details of the algorithm. The first goal is to adjust f by a known quantity so that the jumps are coherent across all nonplanar curves. At this stage of the algorithm, the edge coherence condition has been tested, and it is assumed to be valid.

We show how to make the jumps of f coherent across irreducible nonplanar components of an intersection of a cone with another cone or with a sphere.

Let $f = \sum a_i \chi(A_i)$, and let C be an irreducible nonplanar curve for which the edge coherence condition holds. We assume that it is congruent to some boundary curve C_j of f. Fix an irreducible quadratic function f_2 whose zero set contains C. Partition C into interval I_r as was done in the construction of $D_C(f)$. Automorphisms of C permute the intervals I_r. Let $\mathcal{I}$ be a minimal set of intervals in the sense that every interval I_r is congruent to exactly one element of $\mathcal{I}$.

Let $A = A_i$ be a constituent of f that has C_j as a boundary curve. Let f_1 be an irreducible quadratic function whose zero set contains C_j and such that the jumps through $f_1 = 0$ are not coherent across C_j along some interval I'' of C_j. There is a unique $I \in \mathcal{I}$ and unique congruence $T : C \to C_j$ such that $I'' = T(I)$.

First we handle the cases when $T_* f_2$ and f_1 are not proportional functions.

Assume also that $f_1 = 0$ defines a cone. Recall that each interval $I'' = T(I)$ was defined on some small chart, which is a rectangle on the cone, possibly cut into smaller pieces by planes. Each rectangle on the surface of the cone uniquely determines a wedge of a frustum. The planes cutting the rectangle into a chart cut the frustum into smaller pieces. One of those pieces F corresponds to the interval $T(I)$. The function $T_* f_2$ cuts F into two pieces

$$F_{\pm} = \{x \in F \mid \pm(T_* f_2)(x) > 0\}.$$

The fresh boundary from this cut meets the chart along $T(I)$.

When $f_1 = 0$ defines a sphere, the argument is almost the same. In this case the charts lie on the surface of a sphere. The chart is a spherical triangle, which is the boundary of a uniquely determined solid triangle F. The function $T_* f_2$ cuts F into two pieces. In both cases (both cone and sphere), the pieces $F_{\pm}$ belong to $\mathcal{A}$. The only nonplanar edge on these pieces is the interval I'' on C_j.

There are unique constants $b_{\pm}$ such that

$$g = a_i \chi(A_i) + b_+ \chi(F_+) + b_- \chi(F_-)$$

has

$$J(g, f_1, \pm T_* f_2, C_j, I'') = 0.$$

In particular, with this choice of constants, the jumps are coherent through $f_1 = 0$ across I''. We replace f with the function

$$f + b_+(\chi(F_+) - \chi(T^{-1}F_+)) + b_-(\chi(F_-) - \chi(T^{-1}F_-)).$$

It has several important properties. It is equivalent to f. It satisfies the edge coherence condition if and only if the function f does. The incoherent jump across C_j along $T(I)$ has been translated by the rigid motion T to an incoherent jump across C along I.

Call this equivalent function f. Repeat this procedure until we have moved all incoherent jumps to intervals $I \in \mathcal{I}$, except possibly when f_1 is proportional to $T_* f_2$. If we take a small loop around an interval I'', we get a jump in value in the function f each time we cross a boundary surface. As we make a full loop, we return to the original value of the function f. Thus, the sum of the jumps as we complete a loop is zero. Once all the jumps around C_j are zero, except those along $T_* f_2 = 0$, then the zero sum condition forces the jumps along $T_* f_2 = 0$ to be coherent. Thus, coherence of this final surface is automatic, and we find that the only incoherent jumps are confined to interval $I \in \mathcal{I}$.

By construction, there are no congruences between different intervals in $\mathcal{I}$. We have 'used up' all the congruences and automorphisms. This implies that the edge coherence condition for $D_C(f)$ yields that the jumps of f are coherent through every f_1 across every interval I''.

This completes this stage of the algorithm: we have replaced f with an equivalent function that has the property that all jumps are coherent through irreducible nonplanar curves.

8   Planar Intersections

At this point in the algorithm, every jump that is not coherent is across a planar curve. The final steps are to eliminate spherical boundaries and to eliminate conical boundaries. What remains will be a polyhedron, which can be triangulated into tetrahedra, which are primitive regions.

8.1   Spherical Surfaces

In this step we assume that all boundary curves are planar. For quadratic regions, this implies that the curves are lines or conic sections.

We work through the spherical surfaces in groups according to the radius of the sphere, starting with the largest radius. In this step of the algorithm there are no necessary conditions for equidecomposability. The spherical surfaces can always be eliminated.

When one of the boundary surfaces is a sphere, the intersections are always circles. Thus, the spherical part of the boundary of a quadratic region forms a spherical petal figure. We have seen in Section 2.1 how to decompose a spherical petal figure into spherical caps and spherical triangles. Corresponding to this decomposition of the spherical petal figure is a decomposition of the solid ball into wedges of solid caps and solid triangles. These are primitives. Thus, we can always eliminate a given spherical surface.

The boundary of the wedge of a solid cap consists of three planar surfaces and a spherical surface. The boundary of the solid triangle also consists of three planar surfaces and a spherical surface. In this process, we may introduce new jumps along lines C given by the intersection of two planes. These will be handled at a later stage of the algorithm.

8.2 Conic Sections

At this point of the algorithm we assume that every jump occurs along is bounded by planar and conical surfaces, and that every boundary curve is planar (a line or conic section).

The next step is to produce coherence of jumps along cones across conic sections (other than lines and circles). We will deal with lines and circles later. This part of the algorithm is similar to the elimination of nonplanar curves.

We eliminate boundary curves that are not lines and circles. We have a necessary condition that must be satisfied. If this condition is not satisfied, then equidecomposability fails, and the algorithm terminates.

(Conic Section Coherence Condition) For every conic section C, other than circles and lines, every interval I_r of C constructed as above, and every nonzero quadratic function f_1 defining a cone that contains C, the function $D_C(f)$ has jumps through $f_1 = 0$ that are coherent across C along I_r.

The procedure is essentially identical to the process of eliminating nonplanar curves on cones. We fix a conic section C that satisfies the conic section coherence condition. We let F be a suitable frustum, which we cut into two pieces $F_\pm$ by a plane that meets the conic boundary along an interval $I'' \subset C_j$, for some conic section C_j that is congruent to C. The pieces $F_\pm$ lie in $\mathcal{A}$. The edges of the pieces $F_\pm$ consist of $I'' \subset C_j$, and linear and circular segments. By means of a congruence $T : C \to C_j$, we transport the jumps so that they occur along an interval of C, rather than an interval of C_j. Once all the jumps lie along intervals of C, and once we have 'used up' the congruences and automorphisms, the conic section coherence condition for conic section implies coherence of jumps.

8.3 Circles

At this stage of the algorithm we assume that all boundary curves are lines and circles. These lines and circles form isothermal coordinates on each cone. Using these lines and circles, we break the surface into curvilinear rectangles on the cone. Each such rectangle is the conic surface of a uniquely defined wedge of a frustum, which is a primitive region. Subtracting off these frustums, we obtain a region in which the jump across each conical surface is zero. In other words, they no longer form part of the boundary of the quadratic regions. Thus, after subtracting the frustums, we are left with a polyhedron.

8.4   Polyhedra

As we mentioned above, once all spherical and conical surfaces have been eliminated, we are left with a polyhedron. This can always be triangulated into tetrahedra, which are primitives. Thus, the algorithm is complete.

9   Example

In this section we show an explicit example of a volume computed by this algorithm. The region we consider occurs many times in the proof of the Kepler conjecture [3]. It is called a *quoin*.

Let a, c, t be constants with $a < c$ and $0 < t$. We define the following quadratic region $Q(a, c, t)$:

$$\{(x, y, z) \mid z > 0,\ a < y < tx,\ x^2 + y^2 + z^2 < c^2\}.$$

Note that if $Q(a, c, t)$ is nonempty, we have

$$(a/t)^2 + a^2 + 0^2 < x^2 + y^2 + z^2 < c^2.$$

This condition implies that $t = a/\sqrt{b^2 - a^2}$ with $a < b < c$. We assume that this condition holds for some b. The volume $q(a, c, t)$ is then given explicitly as follows:

$$\begin{aligned}
6\, q(a, c, t) = \ &(a + 2c)(c - a)^2 \arctan(e) \\
&+ a(b^2 - a^2)e \\
&- 4c^3 \arctan(e(b - a)/(b + c)),
\end{aligned} \tag{1}$$

where $e \geq 0$ is given by $e^2(b^2 - a^2) = (c^2 - b^2)$.

This formula is obtained by applying the algorithm to the given quadratic region. All the boundary curves are planar. In fact, every curve is a circular arc or a line segment. No surfaces are cones. Thus, the volume is computed by a particularly simple application of the algorithm.

We see that the intersection of $Q(a, b, t)$ with the sphere of radius c is a spherical petal figure. After subtracting off the contribution from the petal figure, we are left with a tetrahedron. This leads to the given formula for volume.

10   Applications to Flyspeck

In 1998, Sam Ferguson and I gave a proof of the Kepler Conjecture, which asserts that no packing of congruent balls has density greater than the face-centered cubic packing. The proof relies on a significant number of computer calculations.

The refereeing process took several years. As a result of the difficulties in checking the correctness of the Kepler conjecture, I have become interested in formal theorem proving, as a way of checking complex computer proofs. In 2003, I announced a project called Flyspeck, whose purpose is to give a completely formal proof of the Kepler Conjecture [4]. The name 'Flyspeck' comes as an expansion of the acronym 'F*P*K', which stands for the 'Formal Proof of Kepler.'

In addition to the computer part of the proof, the proof of the Kepler Conjecture involves nearly 300 pages of traditional mathematical arguments. The Flyspeck project aims to formalize the traditional mathematical portions of the proof as well. The non-computer parts of the project are not nearly so far along. In my view, a major impediment to completing this part of this project is a lack of modularity in the design of the original proof.

A significant part of these 300 pages consists of volume calculations of quadratic regions. In the original proof, these were all obtained by hand, using a variety of techniques. Every one of the volume calculations falls within the scope of the algorithm of this paper. (They can all be expressed as a linear combination of primitives.) As a result of this paper, these calculations can be entirely automated.

References

1. Basu, S., Pollack, R., Roy, M.-F.: Algorithms in real algebraic geometry. In: Algorithms and Computation in Mathematics, Second edition, vol. 10, Springer, Berlin (2006)
2. Berberich, E.: Exact Arrangements of Quadric Intersection Curves, Master's thesis, Saarbrücken (2004)
3. Ferguson, S., Hales, T.: The Kepler Conjecture. Disc. and Comp. Geom. 36(1), 1–269 (2006)
4. Hales, T.: The Flyspeck Fact Sheet, revised 2007 (2003),
   http://www.math.pitt.edu/~thales/
5. Higbee, F.: The Essentials of Descriptive Geometry. Second edition, John Wiley and Sons, New York (1917)
6. Laczkovich, M.: Equidecomposability and discrepancy; a solution of Tarski's circle-squaring problem. J. Reine Angew. Math. 404, 77–117 (1990)
7. Mishra, B.: Computational Real Algebraic Geometry. In: Goodman, J.E., O'Rourke, J. (eds.) Handbook of Discrete and Computational Geometry, pp. 743–764. CRC Press, Boca Raton FL (1997)

Automatic Verification of Regular Constructions in Dynamic Geometry Systems

Predrag Janičić[1,*] and Pedro Quaresma[2,**]

[1] Faculty of Mathematics, University of Belgrade
Studentski trg 16, 11000 Belgrade, Serbia
`janicic@matf.bg.ac.yu`
[2] Department of Mathematics, University of Coimbra
3001-454 Coimbra, Portugal
`pedro@mat.uc.pt`

Abstract. We present an application of automatic theorem proving (ATP) in the verification of constructions made with dynamic geometry software (DGS). Given a specification language for geometric constructions, we can use its processor to deal with syntactic errors. The processor can also detect semantic errors — situations when, for a given concrete set of geometrical objects, a construction is not possible. However, dynamic geometry tools do not test if, for a given set of geometrical objects, a construction is geometrically sound, i.e., if it is possible in a general case. Using ATP, we can do this last step by verifying the geometric constructions deductively. We have developed a system for the automatic verification of regular constructions (made within DGSs GCLC and Eukleides), using our ATP system, GCLCprover. This gives a real-world application of ATP in dynamic geometry tools.

1 Introduction

Dynamic geometry software (e.g., *Cinderella*, [24,26], *Geometer's Sketchpad*, [10,27] *Cabri*, [14,25]) visualise geometric objects and link formal, axiomatic nature of geometry (most often — Euclidean) with its standard models (e.g., Cartesian model) and corresponding illustrations. The common experience is that dynamic geometry software significantly helps students to acquire knowledge about geometric objects and, more generally, for acquiring mathematical rigour. However, most (if not all) of these programs use only geometric concepts interpreted via concrete instances in Cartesian plane. Namely, a construction is always associated with concrete fixed points (with concrete Cartesian coordinates). In such environments, some constructions (usually by ruler and compass) are illegal (e.g., if they attempt to use the intersection of two parallel lines), but

---

* This work was partially supported by Serbian Ministry of Science and Technology grant 144030. Also, partially supported by the programme POSC, by the Centro International de Matemática (CIM), under the programme "Research in Pairs", while visiting Coimbra University under the Coimbra Group Hospitality Scheme.
** This work was partially supported by programme POSC.

F. Botana and T. Recio (Eds.): ADG 2006, LNAI 4869, pp. 39–51, 2007.
© Springer-Verlag Berlin Heidelberg 2007

the question if such construction is always illegal or it is illegal only for given particular fixed points is left open (if a construction is always possible, we will call it *regular*). Indeed, for answering such question, one has to use deductive reasoning, and not only a semantic check for the special case. Consider one simple example: given (by Cartesian coordinates) three fixed distinct points A, B, C, we can construct a point D as an image of point C in translation $\mathcal{T}_{AB}$; later on, if we try to construct an intersection of lines AC and BD, we will discover that there is no such intersection (since these two lines are parallel). This holds, not for some specific points A, B, C, with D determined as above, but for all triples of points A, B, C. So, in this situation, the user of a geometry tool should get the information that his/her construction is illegal, and moreover, that it is illegal not only for a given special case, but always. In this way, the deductive nature of geometric conjectures and proofs should be linked to the semantic nature of models of geometry and, also, to human intuition and to geometric visualisations.

In the rest of this paper we present our system which addresses the above problem. Our system is implemented within dynamic geometry software GCLC [11] and Eukleides [19,23] and uses a geometry theorem prover, GCLCprover [12], based on the area method [4,5]. Our framework, GeoThms [21,22], is a Web tool that integrates the above components with a repository of theorems related to geometric constructions.

Closely related to our system is Geol — a geometric object-oriented language and a system for geometric computation, reasoning and visualisation [15]. This system focuses on symbolic manipulation of geometric objects (in algebraic form). Regarding handling degeneracy conditions and illegal constructions, there is a *consistency checking* system [16]. When an object is modified, or a new relation among existing object is declared, the system checks if this action is allowed, i.e., if it is consistent with the rest of the construction. This check is reduced to testing if a corresponding algebraic system has solutions in real numbers. Also, related to our system is *Geometry Explorer*, based on the full-angle method [30]. This system provides tight integration of DGS and ATP, and produces human-readable proofs of properties of constructed objects (in LaTeX form). MMP/Geometer also combines features of DGS and ATP, and uses different proving methods, including those generating synthetic, human-readable proofs [8,9]. There are several other systems that in some degree link DGSs with ATPs: *Geometry Expert* (GEX) [7]; GEOTHER [28,29]; *Cinderella* [13,24,26]; *Discover* [2]; *GeoView* [1], and *GeoGebra* [6]. However, none of these systems incorporates a verification system for constructions which provides arguments in the form of synthetic, readable proofs.

Paper Overview. Section 2 briefly discusses geometric constructions, the domain of our system. Section 3 talks about parts of our framework; subsection 3.1 is about dynamic geometry software, especially GCLC and Eukleides; subsection 3.2 is about automated theorem proving in geometry and especially the prover GCLCprover and subsection 3.3 briefly describes the integration of these tools in a Web Geometric Framework. Section 4 presents the verification system and the covered critical constructions. Section 5 presents some examples. Section 6 discusses further work, and in Section 7 we draw final conclusions.

2   Geometric Constructions

For hundreds, or even thousands of years, geometric construction problems have been one of the most attractive parts of geometry and mathematics. A geometric construction is a sequence of specific, primitive construction steps. These primitive construction steps (also called *elementary constructions*) are based on using a *ruler* (or a *straightedge*[1]) and a *compass*, and they are:

– construction (with a *ruler*) of a line such that two given points belong to it;
– construction (with a *ruler*) of a segment connecting two points;
– construction of a point which is an intersection of two lines (if such a point exists);
– construction (with a *compass*) of a circle such that its centre is one given point and such that a second given point belongs to it;
– construction of intersections between a given line and a given circle (if such points exist).

By using this set of primitive constructions, one can define more complex, compound constructions (e.g., construction of a right angle, construction of the midpoint of a line segment, etc.).

The abstract (i.e., formal, axiomatic) nature of geometric objects has to be distinguished from their usual interpretations. A geometric construction is a procedure consisting of abstract steps and it is not a picture. However, for each construction there are its counterparts, its interpretations in the standard Cartesian model.

3   Component Modules of the Automatic Verification System

In this section, we present the building blocks of our automatic verification system for geometric constructions.

3.1   GCLC and Eukleides

GCLC is a tool for teaching and studying mathematics, especially geometry and geometric constructions, and also for storing descriptions of mathematical figures and for producing digital illustrations.[2] GCLC provides support for a

---

[1] The term "straightedge" is sometimes used instead of "ruler" in order to emphasise there are no markings which could be used to make measurements.

[2] GCLC package is freely available from `www.matf.bg.ac.yu/~janicic/gclc/`. The mirrored version is available from EMIS (The European Mathematical Information Service) `www.emis.de/misc/index.html`. There are versions of GCLC for Windows and for Linux.

range of geometric constructions and isometric transformations. In GCLC there is also support for symbolic expressions, second order curves, parametric curves, control structures, etc. GCLC is based on the idea that constructions are formal procedures, rather than drawings. Thus, in GCLC, producing mathematical illustrations is based on "describing figures" rather than of "drawing figures". All figures are described in this spirit, using the GC language. These descriptions directly reflect the mathematical contents, i.e., the meaning of mathematical objects to be presented, and are easily understandable to mathematicians. WinGCLC is the Windows version of GCLC, with a rich graphical interface and it provides a range of additional functionalities to GCLC. It supports interactive work, animations, traces, "watch window" for monitoring values of selected objects, etc. [11].

Eukleides[3] is an Euclidean geometry drawing language. There are two programs related to it. The first is `eukleides`, a processor for describing geometric figures within a (La)TeX document. It can also convert figures in Eukleides format to EPS format. The second is `xeukleides`, a GUI front-end for creating interactive geometric figures. This program can also be used for editing Eukleides code. Eukleides, like GCLC, has been designed to be close to the traditional language of elementary Euclidean geometry. We have developed a tool `euktogclcprover`, that converts Eukleides files to GCLCprover files, enabling the prover to be used with geometric constructions described within both GCLC and Eukleides.

We have developed a XML-based format (and accompanying tools) for representing geometric constructions and proofs. This format enables a suitable rendering of this contents, and also serves as a convenient exchange format between, not only GCLC, Eukleides, and GCLCprover, but other geometric tools as well.

3.2 GCLCprover, an ATP Based on the Area Method

Automated theorem proving in geometry has two major lines of research: synthetic proof style and algebraic proof style (see, for instance, [17] for a survey). Algebraic proof style methods are based on reducing geometry properties to algebraic properties expressed in terms of Cartesian coordinates. These methods are usually very efficient, but the proofs they produce often do not reflect the geometric nature of the problem and, basically, they give only *yes* or *no* conclusions. Synthetic methods attempt to automate traditional geometry proof methods and to produce human-readable proofs.

The area method is a synthetic method providing traditional (not coordinate-based), human-readable proofs [4,5]. The proofs are expressed in terms of higher-level geometric lemmas and expression simplifications. The main idea of the method is to express hypotheses of a theorem using a set of constructive statements, each of them introducing a new point, and to express a conclusion by

---

[3] Eukleides is available from `http://www.eukleides.org`, There are versions for a number of languages. The second author of this paper is responsible for the Portuguese version of Eukleides: EukleidesPT is available from `http://gentzen.mat.uc.pt/~EukleidesPT/`

an equality of expressions in some geometric quantities (e.g., the signed area of a triangle), without referring to Cartesian coordinates. The proof is then based on eliminating (in reverse order) the points introduced, using for that purpose a set of appropriate lemmas. After eliminating all introduced points, the current goal becomes an equality between two expressions in quantities over independent points. If it is trivially true, then the original conjecture was proved valid, if it is trivially false, then the conjecture was proved invalid, otherwise, the conjecture has been neither proved nor disproved. In all stages, different simplifications are applied to the current goal. The method does not have branching, which makes it very efficient. The area method is applicable to a wide range of constructions and a wide range of geometric conjectures. For details of the method and correctness proofs for all simplification steps see [20].

We have implemented GCLCprover, a theorem prover that allows formal deductive reasoning about objects constructed with the help of DGSs. The prover is based on the area method. It produces proofs that are human-readable and with an explicit justification for every proof step. The prover can be used in conjunction with other dynamic geometry tools. Apart from the original implementation by its authors [4,5], we are aware of another two geometry provers based on the area method: one within the Theorema project [3], and one within the system Coq (COQareaMethod) [18].

GCLCprover is tightly integrated with dynamic geometry tools (GCLC and Eukleides). This means that one can use the prover to reason about a a DGS construction (i.e., about objects introduced in it), without changing and adapting it for the deduction process — the user only needs to add the conclusion he/she wants to prove. The geometric constructions made within the DGSs are internally transformed into primitive constructions of the area method, and in some cases, some auxiliary points are introduced.

GCLCprover can prove many complex geometric problems in milliseconds, producing readable proofs (in LATEX or XML form).

3.3   The Geometric Framework

GeoThms[4] is a framework that links dynamic geometry software (GCLC, Eukleides), geometry theorem provers (GCLCprover), and a repository of geometry problems (geoDB). GeoThms provides a Web workbench in the field of constructive problems in Euclidean geometry. Its tight integration with dynamic geometry tools and automatic theorem provers (currently GCLC, Eukleides, GCLCprover and COQareaMethod) and its repository of theorems, figures, and proofs, give the user the possibility to easily browse through a list of geometric problems, their statements, illustrations and proofs. It also provides an interface to the DGS and ATP components, allowing the interactive use of those programs and also supporting the automatic verification of regular constructions performed within the DGSs.

---

[4] GeoThms is accessible from `http://hilbert.mat.uc.pt/~geothms`

4   Integrated Automated Verification System

The system for automated deductive testing whether a construction is regular, is built into the DGSs, GCLC and Eukleides, and uses GCLCprover. While processing the input file (with a description of a geometrical construction), a DGS provides to the built-in theorem prover all construction steps performed (transformed into a suitable form). This system can be switched off or on.

When the main module of GCLC encounters a construction step that cannot be performed (for instance, two identical points do not determine a line), it reports that the step is illegal with respect to a given set of fixed points (at this point, this is only an argument based on semantics, on calculations concerning concrete fixed points), and then it invokes the theorem prover. After that, the prover is ran on the critical conjecture (e.g., it tries to prove that the two points are identical) and, if successful, it reports that the construction step is always illegal/impossible. This is a result of a deduction process based on formal description of constructions, not on coordinates of the concrete points involved.

We point out that the "errors" that our deduction system detects and reports about are substantially different from syntax errors detected by the parsing modules of DGSs. Syntax errors are usually simple, local, must be eliminated from the description of the construction, and are not related to any deeper underlying geometrical knowledge. On the other hand, an illegal construction detected by our system signals the user to reconsider the whole of the construction, and claims that the construction is impossible no matter how the fixed points were selected. From a semantical point of view, we can eliminate some of the errors that our system reports about: for instance, if we use homogeneous coordinates, we could treat intersections of lines uniformly and there would be no exception for parallel lines. However, we don't want our tool to *avoid* errors within constructions, we want to explore the properties that are deeply related to the intended construction and to guide the user through the construction process. This approach reveals properties of Euclidean constructions, therefore it also has an educational role.

Realm. Our automatic verification deductive-check system currently covers the following critical constructions:

- constructing a line given two points (error if the two points are identical);
- constructing an intersection of two lines (error if the two lines are parallel);
- constructing a segment-bisector, given its two endpoints (error if the two points are identical);
- constructing an angle-bisector of the angle determined by three points A, B, C (error if A and B, or C and B are identical);
- calculating an angle determined by three points A, B, C (error if A and B, or C and B are identical);

Geometric objects that are subject to deductive verification have to be made within the DGSs using the following primitives:

- introducing a new point;
- constructing a line given two points;

- determining the intersection of two lines;
- constructing the midpoint of a segment;
- constructing the segment bisector;
- constructing the line passing through a given point, perpendicular to a given line;
- constructing the foot from a point to a given line;
- constructing the line passing through a given point, parallel to a given line;
- constructing the image of a point in a given translation;
- constructing the image of a point in a given scaling transformation;
- selecting a random point on a given line.

which are internally transformed into primitive constructions of the area method. For more details about this transformation see [20].

It is worth pointing out that although GCLC and Eukleides have support for a large number of constructions, only few of them can be illegal. The above list of critical constructions almost exhaust them. The only possible illegal constructions which are not covered by the current version of our system are constructions of intersection points of a circle and a line, and of two circles. Corresponding geometric conjectures cannot be generally handled by the area method and GCLCprover. In our future work, we will consider extending our system by additional deduction methods that can also cover this sort of constructions.

5   Worked Examples

In this section we give several examples for which our system can deductively test if they are regular. There is also one example that is out of the scope of the current version of our system.

5.1   Example 1

Consider the example discussed in Section 1: given three fixed distinct points A, B, C, let us construct a point D as an image of the point C in translation T_{AB}; draw lines AB and CD (denoted p and q) and label all the points. The GCLCcode for this construction and the corresponding illustration, are shown in Figure 1.

If we attempt to construct a point X as the intersection of the lines p and q (by adding the command `intersec X p q` at the end of the code given in Figure 1), we will get the following message:

```
Error 14: Run-time error: Bad definition.
Can not determine intersection. (Line: 18, position: 10)
File not processed.
```

This information is semantic-based, it is true for the given particular points A, B, C, i.e., for these three particular points, the lines p and q are parallel.

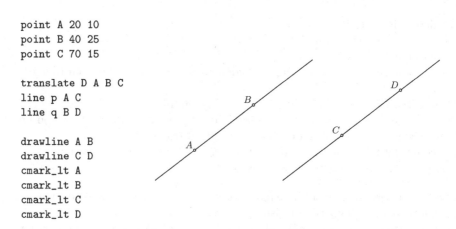

```
point A 20 10
point B 40 25
point C 70 15

translate D A B C
line p A C
line q B D

drawline A B
drawline C D
cmark_lt A
cmark_lt B
cmark_lt C
cmark_lt D
```

Fig. 1. GCLC code and the corresponding illustration for the example with parallel lines

However, if our deductive-check system is turned on, we will also get additional, much deeper information:

```
Deduction check invoked: the property that led to the error is
tested for validity.

Total number of proof steps:        18

Time spent by the prover: 0.001 seconds
The conjecture successfully proved - the critical property always holds.
The prover output is written in the file error-proof.tex.
```

This means that it was proved that lines p and q never intersect, so this construction is always illegal. The proof of this fact is generated by the prover GCLCprover and the proof outline is given in Figure 2. Note that the condition $p \| q$ is equivalent to the condition that the areas of triangles ABD and CBD are equal.

5.2   Example 2

Consider the example given in Figure 3. This example is very similar to the previous one, the only difference is in the way point D is determined. In both cases point D gets the same Cartesian coordinates. However, in the first example, D is determined by a construction based on the points A, B, C. In contrast, in the second example, point D is determined by Cartesian coordinates, independently from the points A, B, C.

This time, it is not possible to deduce that this construction is always illegal:

```
Run-time error: Bad definition. Can not determine intersection.
(Line: 16, position: 10)
```

(1)	$S_{ABD} = S_{CBD}$	,	by the statement
(2)	$(S_{ABC} + (1 \cdot (S_{ABB} + (-1 \cdot S_{ABA})))) = S_{CBD}$	,	by Lemma 29 (point D eliminated)
(3)	$(S_{ABC} + (1 \cdot (0 + (-1 \cdot 0)))) = S_{CBD}$	,	by geometric simplifications
(4)	$S_{ABC} = S_{CBD}$	,	by algebraic simplifications
(5)	$S_{ABC} = (S_{CBC} + (1 \cdot (S_{CBB} + (-1 \cdot S_{CBA}))))$	,	by Lemma 29 (point D eliminated)
(6)	$S_{ABC} = (0 + (1 \cdot (0 + (-1 \cdot (-1 \cdot S_{ABC})))))$	,	by geometric simplifications
(7)	$0 = 0$	,	by algebraic simplifications

Q.E.D.

Fig. 2. Proof of the critical property for example 5.1

```
point A 20 10
point B 40 25
point C 70 15
point D 90 30

line p A C
line q B D

drawline A B
drawline C D
cmark_lt A
cmark_lt B
cmark_lt C
cmark_lt D

intersec X p q
```

Fig. 3. GCLC code for example with parallel lines and the point D given by Cartesian coordinates

```
Deduction check invoked: the property that led to the error will
be tested for validity.

The conjecture not proved - the critical property does not
always hold.
The prover output is written in the file error-proof.tex.
```

5.3   Example 3

Consider a more elaborate example (see Figure 4): let O_1 and O_2 be the pairwise intersections of the side-bisectors of triangle ABC. These two points are

```
point A 30 10
point B 70 10
point C 60 45

med a B C
med b A C
med c B A

intersec O_1 a b
intersec O_2 a c

drawsegment A B
drawsegment A C
drawsegment B C
cmark_b A
cmark_b B
cmark_t C
cmark_lb O_1
cmark_rb O_2

line p O_1 O_2
```

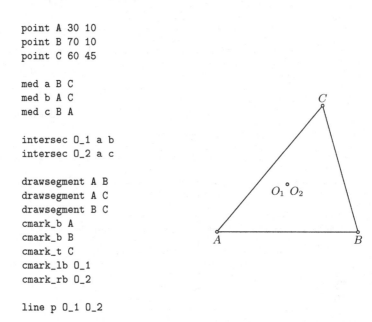

Fig. 4. GCLC code and the corresponding illustration for the example with two identical points

always identical, so the construction of a line p determined by these two points is not possible. When the system encounters this construction step, it invokes the prover which successfully proves that this step is always illegal.

```
point A 20 25
point O 60 25
point X 60 40

circle k O X
midpoint M O A
circle l M A

intersec2 T_1 T_2 k l
line t A T_2
intersec2 P_1 P_2 k t

cmark_t A
cmark_t O
cmark_lb T_2
drawcircle k
drawline t

line p P_1 P_2
```

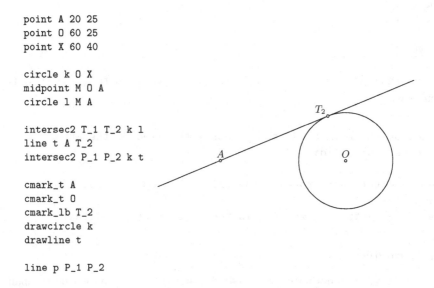

Fig. 5. GCLC code and the corresponding illustration for the example with tangent

5.4   Example 4

In the example shown in Figure 5, line t contains point A and is the tangent to circle k. The constructions is as follows: let k be the circle with centre O passing through the point X; let M be the midpoint of the segment OA; let l be the circle with centre M and passing through the point A; let T_1 and T_2 be the intersection points of the circles k and l. Since T_2 belongs to l, it holds that the angle AT_2O is a right angle. Since T_2 belongs to k, it follows that T_2 belongs to the tangent from A to k. Let us denote by t the tangent AT_2 from A to k. Since t is a tangent, its two intersection points with k, P_1 and P_2, are identical. Therefore, P_1 and P_2 do not determine a line, which is relevant for the construction step `line p P_1 P_2`. This is detected by the main construction module (for the given, specific points), but the prover fails to prove it (because of the realm of the area method, see Section 4).

6   Further Work

Our verification system checks if a construction is illegal, i.e., if it is always impossible (no matter how the starting points were selected). While the construction (with its Cartesian counterpart) is being performed, the verification system is invoked, when and if, for a given set of fixed points, a construction step cannot be performed. We can extend our system so it can also invoke deduction checks for all construction steps even if there is no semantic error encountered. This would give verified regular constructions — constructions that can always be performed (provided the fixed points meet some conditions). However, such system would be time consuming (as it would run the theorem prover for each construction step).

We are planning to further improve the underlying deducting module, and to implement other geometry theorem provers, covering constructions that are out of the realm of the current system.

Also, we are planning to develop a support for guided step-by-step constructions. Such a tutoring system would control each user step, both in syntax, semantics, and deductive terms, and would serve as a teaching assistant for studying geometry.

7   Conclusions

In this paper, we presented our system for automatic verification of constructions. It provides a deep argument why a certain construction is not regular and it gives a new power to dynamic geometry tools. The system is used within dynamic geometry tools GCLC and Eukleides, and in the wider context of our publicly available geometric framework GeoThms. The underlying deduction module is based on the area method for Euclidean geometry. For future work, we are planning to implement some other geometry prover, and to further extend the realm of our system. We are also planning to extend the system so it can be used as a tutor for studying geometry.

Acknowledgements. We are grateful to Reinhard Kahle for an inspiring discussion about topics presented in this paper.

References

1. Bertot, Y., Guilhot, F., Pottier, L.: Visualizing geometrical statements with GeoView. Electr. Notes Theor. Comput. Sci. 103, 49–65 (2004)
2. Botana, F., Valcarce, J.L.: A dynamic-symbolic interface for geometric theorem discovery. Computers and Education 38, 21–35 (2002)
3. Buchberger, B., Craciun, A., Jebelean, T., Kovacs, L., Kutsia, T., Nakagawa, K., Piroi, F., Popov, N., Robu, J., Rosenkranz, M., Windsteiger, W.: Theorema: Towards computer-aided mathematical theory exploration. Journal of Applied Logic 4(4), 470–504 (2006)
4. Chou, S.-C., Gao, X.-S., Zhang, J.-Z.: Automated production of traditional proofs for constructive geometry theorems. In: Vardi, M. (ed.) LICS. Proceedings of the Eighth Annual IEEE Symposium on Logic in Computer Science, pp. 48–56. IEEE Computer Society Press, Los Alamitos (1993)
5. Chou, S.-C., Gao, X.-S., Zhang, J.-Z.: Automated generation of readable proofs with geometric invariants, I. multiple and shortest proof generation. Journal of Automated Reasoning 17, 325–347 (1996)
6. Fuchs, K., Hohenwarter, M.: Combination of dynamic geometry, algebra and calculus in the software system geogebra. In: Computer Algebra Systems and Dynamic Geometry Systems in Mathematics Teaching Conference 2004, Pecs, Hungary, pp. 128–133 (2004)
7. Gao, X.-S.: GEX, http://www.mmrc.iss.ac.cn/~xgao/software.html
8. Gao, X.-S., Lin, Q.: MMP/Geometer - a software package for automated geometric reasoning. In: Winkler, F. (ed.) ADG 2002. LNCS (LNAI), vol. 2930, pp. 44–66. Springer, Heidelberg (2004)
9. Gao, X.S., Lin, Q.: MMP/Geometer,
   http://www.mmrc.iss.ac.cn/~xgao/software.html
10. Jackiw, N.: The Geometer's Sketchpad v4.0. In: Emeryville: Key Curriculum Press (2001)
11. Janičić, P.: GCLC — a tool for constructive euclidean geometry and more than that. In: Iglesias, A., Takayama, N. (eds.) ICMS 2006. LNCS, vol. 4151, pp. 58–73. Springer, Heidelberg (2006)
12. Janičić, P., Quaresma, P.: System Description: GCLCprover + GeoThms. In: Furbach, U., Shankar, N. (eds.) IJCAR 2006. LNCS (LNAI), vol. 4130, pp. 145–150. Springer, Heidelberg (2006)
13. Kortenkamp, U., Richter-Gebert, J.: Using automatic theorem proving to improve the usability of geometry software. In: Procedings of the Mathematical User-Interfaces Workshop 2004 (2004)
14. Laborde, J.M., Strässer, R.: Cabri-géomètre: A microworld of geometry guided discovery learning. International reviews on mathematical education- Zentralblatt fuer didaktik der mathematik 90(5), 171–177 (1990)
15. Liang, T., Wang, D.: Towards a geometric-object-oriented language. In: Hong, H., Wang, D. (eds.) ADG 2004. LNCS (LNAI), vol. 3763, pp. 130–155. Springer, Heidelberg (2006)
16. Liang, T., Wang, D.: Geometric constraint handling in Gool. In: Automated Deduction in Geometry, pp. 66–73 (2006)

17. Matsuda, N., Vanlehn, K.: Gramy: A geometry theorem prover capable of construction. Journal of Automated Reasoning 32, 3–33 (2004)
18. Narboux, J.: A decision procedure for geometry in Coq. In: Slind, K., Bunker, A., Gopalakrishnan, G.C. (eds.) TPHOLs 2004. LNCS, vol. 3223, pp. 225–240. Springer, Heidelberg (2004)
19. Obrecht, C.: Eukleides, http://www.eukleides.org/
20. Quaresma, P., Janičić, P.: Framework for constructive geometry (based on the area method). Technical Report 2006/001, Centre for Informatics and Systems of the University of Coimbra (2006)
21. Quaresma, P., Janičić, P.: GeoThms - geometry framework. Technical Report 2006/002, Centre for Informatics and Systems of the University of Coimbra (2006)
22. Quaresma, P., Janičić, P.: Integrating dynamic geometry software, deduction systems, and theorem repositories. In: Borwein, J.M., Farmer, W.M. (eds.) MKM 2006. LNCS (LNAI), vol. 4108, pp. 280–294. Springer, Heidelberg (2006)
23. Quaresma, P., Pereira, A.: Visualização de construções geométricas. Gazeta de Matemática 151, 38–41 (2006)
24. Richter-Gebert, J., Kortenkamp, U.: The Interactive Geometry Software Cinderella. Springer, Heidelberg (1999)
25. Cabri Web site, http://www.cabri.com
26. Cinderella Web site, http://www.cinderella.de
27. Geometer's Sketchpad Web site, http://www.keypress.com/sketchpad/
28. Wang, D.: Geother (geometry theorem prover),
    http://www-calfor.lip6.fr/~wang/GEOTHER/
29. Wang, D.: GEOTHER 1.1: Handling and proving geometric theorems automatically. In: Winkler, F. (ed.) ADG 2002. LNCS (LNAI), vol. 2930, pp. 194–215. Springer, Heidelberg (2004)
30. Wilson, S., Fleuriot, J.: Combining dynamic geometry, automated geometry theorem proving and diagrammatic proofs. In: Proceedings of UITP 2005, Edinburgh, UK (April 2005)

Recognition of Computationally Constructed Loci

Peter Lebmeir and Jürgen Richter-Gebert

Technical University of Munich, Department of Mathematics,
Boltzmannstr. 3, 85748 Garching, Germany
lebmeir@ma.tum.de,
richter@ma.tum.de
http://www-m10.ma.tum.de/bin/view/Lehrstuhl/PeterLebmeir
http://www-m10.ma.tum.de/bin/view/Lehrstuhl/RichterGebert

Abstract. We propose an algorithm for automated recognition of computationally constructed curves and discuss several aspects of the recognition problem. Recognizing loci means determining a single implicit polynomial equation and geometric invariants, characterizing an algebraic curve which is given by a discrete set of sample points. Starting with these discrete samples, arising for example from a geometric ruler and compass construction, an eigenvalue analysis of a matrix derived from the data leads to proposed curve parameters. Utilizing the construction itself, with its free and dependent geometric elements, further specifications of the type of constructed curves under genericity assumptions are made. This is done by a second eigenvalue analysis of parameters of several generically generated curves.

1   Introduction

The generation of loci is one of the central applications in todays computer geometry programs. In abstract terms a *locus* consists of all locations that a specific point of a geometric configuration can take, while one parameter of the configuration may vary. For instance the locus of all points that are at a fixed distance from a fixed point is simply a circle. Typically the locus data generated by a geometry program does not consist of a symbolic description of the locus. Rather than that a more or less dense collection of sample points on the locus is generated. In many cases the user of such a program can identify the underlying curve visually by simply looking at it or by a pre-knowledge of the underlying construction or by a combination of both. However, for several applications (like automatic assistance, geometric expert systems, etc.) it is highly desirable to recognize these curves automatically. After a geometric construction for a locus is specified such a recognition procedure can be carried out on two different levels. First, one is interested in which locus is generated for a *concrete* instance (i.e. position of free elements) of the construction. Second, one is interested in invariants of *all* the loci that can be generated by a specific construction. We refer to this set as the *type* of the locus. Very often in the literature *types of*

F. Botana and T. Recio (Eds.): ADG 2006, LNAI 4869, pp. 52–67, 2007.

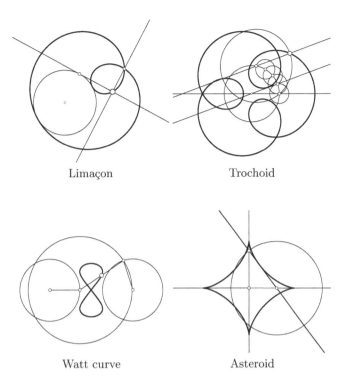

Limaçon   Trochoid

Watt curve   Asteroid

Fig. 1. Geometric constructions of loci; each curve can be interpreted as zero set of an algebraic equation

loci (which are algebraic curves in our setup) are closely related to *names* of the corresponding curves like *cardioide, limaçon, lemniscate, Watt curve, conic,* etc. One of the ultimate goals of a locus recognition algorithm would be an output like:

> "You constructed a limaçon with parameters $r = 0.54$ and $s = 0.72$. A limaçon is given by the equation $(x^2 + y^2 - 2rx)^2 = s^2(x^2 + y^2)$. Your construction will generically generate a limaçon."

If one treats the geometric construction as a black box which takes positions of the free elements of a construction as input and produces sample points of the generated locus as output, the above demands imply the necessity of a curve recognition algorithm based on discrete sets of sample points.

In our constructions we only allow primitive operations coming from ruler and compass operations. In this case a locus turns out to be contained in a zero set of a polynomial equation for any specific instance of a construction. Fixing the free construction elements in a way that only one degree of freedom is left, a dependent point is restricted to an algebraic curve. The locus data is given by a discrete set of sample points $\mathcal{P}$ on this curve, constructed for example via a dynamic geometry program. In general, these programs trace a single point and thus return a set $\mathcal{P}$ contained in a single real branch of the zero set of an algebraic equation. Therefore we presume $\mathcal{P}$ to be of that kind.

In our case, curve recognition for a specific instance of a construction means extracting a single algebraic *equation* of minimal degree from a discrete point set $\mathcal{P}$ and determining its degree. In a second step we focus on the set $\mathcal{B}$ of curves obtainable by *all* instances of a specific construction – the *type* of the curve. We presume that all coordinates of dependent elements in a construction are analytic functions in the free parameters of the free elements of a construction. Under this assumption, an examination of the degree of the curves in $\mathcal{B}$ and geometric invariant extraction is possible. In this way we can, at least for simple curves, associate a *name* with $\mathcal{B}$, characterizing the type of the contained curves. In particular the minimal degree itself turns out to be an invariant property.

Recognizing implicit multivariate polynomials has been investigated also by other research groups most often in different contexts and with different side constraints. A genuine feature of the setup treated in this article is that the sample points usually come with a high arithmetic precision and that it is known in advance that they belong to an algebraic curve with an a priori upper degree bound (usually the real degree will be much lower than this bound). This is in contrast to related research work in computer vision or patter recognition where one is interested to approximate camera pictures of geometric shapes by algebraic curves in order to recognize these shapes. There the shape data does not a priori belong to an algebraic curve and one is only interested in relatively roughly approximating curves of fixed degree (compare for instance [6],[9],[10]).

Another related setup was treated and implemented in the dynamic geometry program Cabri Géomètre [3]. There a first locus recognition algorithm was implemented. Unfortunately no algorithmic details are available and practical experiments show that the algorithm used there is relatively instable in particular under rigid transformations. (We will deal with this particular issue later in Section 5.2).

In this article we introduce a concept of locus recognition, that deals with the construction on two different levels. On both levels the construction itself is treated as a black box that generates sample points on the locus for specific parameters of the free elements of the construction. In the first step we propose an algorithm that takes the sample points of one locus as sole input data and reconstructs the parameters and degree of the underlying algebraic curve for this specific instance. In a second step we allow the free construction elements to vary and thus produce a whole collection of sample point sets $\mathcal{P}_1, \mathcal{P}_2, \ldots$ each of which represents locus data of the same type. Randomization techniques are then used to finally extract common features of these loci and determine the type of the locus (with high probability).

2   Computationally Constructed Curves

We use ruler and compass constructions as starting point for our investigations. They provide us with locus data of high arithmetic precision. In principle it does not matter where discrete sample point sets come from, as long as they are very close to an algebraic curve.

Dynamic geometry programs facilitate ruler and compass constructions in a plane: Elementary construction steps consist of placing free points in the plane, joining two points by a line, constructing circles by midpoint and perimeter and intersecting lines with lines, lines with circles or circles with circles. Furthermore, points can be restricted to already constructed elements like a line or a circle. We will call such points *semi-free* as they have only one degree of freedom. Lines (circles) are zero sets of linear (quadratic) equations and can be computationally represented by the parameters of these equations. Semi-free points can be described by a one-parametric solution manifold of an equation. All-in-all (for details see [1]) any instance of a geometric construction corresponds to the solution set of a finite set of polynomial equations in the parameters generated by the free and semi-free elements. Allowing only finitely many ruler and compass construction steps, every position of a dependent point can be described by algebraic equations depending on the position of the free points.

While moving a free point in a dynamic geometry program, one can watch the dependent points move consistently with the construction. Fixing all (semi-) free elements except for one semi-free point in a construction, a single dependent point p gets restricted to a zero set of an algebraic equation. Tracing p with a dynamic geometry program under the movement of a single semi-free point (a *mover*), a single branch B of an algebraic curve is revealed. The algebraic curve can be described by a polynomial equation $b(p) = 0$ of minimal degree. B is the set of locations of the traced point p. Unless otherwise stated all points and curves are assumed to be given in homogeneous coordinates. Under these assumptions a constructed branch B is contained in the zero set $Z(b)$ of a homogeneous polynomial function $b : \mathbb{RP}^2 \to \mathbb{R}$ of minimal degree d_b:

$$
Z(b) = \left\{ \begin{pmatrix} x_1 \\ x_2 \\ x_3 \end{pmatrix} \in \mathbb{RP}^2 \;\middle|\; b(x_1, x_2, x_3) = \sum_{i+j+k=d_b} \beta_{i,j,k} \cdot x_1^i x_2^j x_3^k = 0 \right\} \quad (1)
$$

From a construction in a dynamic geometry program we get a discrete set $\mathcal{P}$ containing points lying almost exactly on a branch B of some $Z(b)$. Using algebraic methods, we could in principle determine a corresponding curve of very high degree by the following approach: In a first step each elementary construction step is described as a polynomial relation in the coordinates of geometric elements. In a second step elimination techniques (for instance based on Gröbner Bases or Ritt's algebraic decomposition method) are used to find an implicit representation of the possible position of the locus generating point. These type of approach to curve recognition is purely symbolic but it significantly suffers from combinatorial explosion of the algebraic structures. Even for small constructions one could in general not hope for results in reasonable computation time. In contrast we use the construction as a "black box" and consider only the sample point data of the computationally constructed branch B. Our aim is to derive its describing algebraic equation b of minimal degree. Now our driving questions are:

1. Which plane algebraic curve of minimal degree is "reasonably" fitting the numerical locus data? (Which b fits $\mathcal{P}$ "reasonably" and what is d_b?)
2. Regarding a ruler and compass construction, which type of plane algebraic curves are constructed? (What is the degree d_b of almost all curves B in $\mathcal{B}$ and what name characterizes $\mathcal{B}$?)

Clearly the problem needs some mathematic modeling which preserves quality criteria. The next section will deal with this issue. Additionally the first part of our work, i.e. the parameter extraction, can be applied to any curve or even surface recognition problem given sufficiently many points of sufficiently high precession.

3   Curve Recognition

Given a finite set of data points $\mathcal{P} = \{p_1, p_2, ..., p_m\}$, e.g. a set of samples in homogeneous coordinates on a locus from a ruler and compass construction. The problem of fitting an algebraic curve B to the data set $\mathcal{P}$ is usually cast as minimizing the mean square distance

$$\frac{1}{m} \sum_{i=1}^{m} dist(p_i, Z(b))^2 \tag{2}$$

from the data points to the curve (see [5]). This is a function in the coefficients β of the polynomial b, where $dist(p, Z(b))$ denotes the Euclidean distance from a point p to the zero set $Z(b)$.

Unfortunately, there is in general no closed form for $dist$. In principle $dist$ could be approximated by iterative calculations. However, they turn out to be expensive and very clumsy for the needed optimization procedure. Therefore we will use other approximation. Without any numerical noise $b(p) = 0$ for all $p \in Z(b)$ the first choice to replace (2) is

$$\frac{1}{m} \sum_{i=1}^{m} b(p_i)^2. \tag{3}$$

We will use this formula as central ansatz for modeling the distance of curve and sample points. This is adequate since our sample points are given with high arithmetic precision.

We now will describe how the coefficients of b that minimize (3) can be calculated. For this let $T_1(x), T_2(x), ..., T_k(x)$ be a basis of the linear space of all homogeneous polynomials of degree d in three variables. If $d = 2$ one could take, for instance, the monomial basis

$$T_1(x) = x_1^2, T_2(x) = x_1 x_2, T_3(x) = x_1 x_3, T_4(x) = x_2^2, T_5(x) = x_2 x_3, T_6(x) = x_3^2.$$

In general we have $k = \frac{1}{2}(d+1)(d+2)$ basis polynomials. Furthermore let

$$\tau_d = (T_1, T_2, \ldots, T_k) : \mathbb{RP}^2 \to \mathbb{R}^k$$

be the function that associates a point $p \in \mathbb{RP}^2$ with its evaluation $\tau_d(p) \in \mathbb{R}^k$ of all the basis functions of degree d.

Assuming we have a discrete set of n points $\mathcal{P} = \{p_1, \ldots, p_n\}$ from an underlying construction and assuming we know d_b, the degree of the computationally constructed branch B, we can fix τ_d for polynomials of degree d_b. Furthermore, we can form an $n \times k$ matrix $P := P_{d,\mathcal{P}}$ with row vectors $\tau_d(p_i)$ of τ_d-transformed points $p_i \in \mathcal{P}$. Thus minimizing (3) is equivalent to minimizing $\|P\beta\|_2$, with the unknown parameter vector β. So far the minimum can easily be obtained by setting $\beta = (0, \ldots, 0)$. This comes from the homogenhomogeneityicity of the problem setup and the fact that β and $\lambda \cdot \beta$ for $\lambda \in \mathbb{R} \backslash \{0\}$ define the same curve. We overcome this problem by forcing additional side-constraints on β and require $\|\beta\|_2 = 1$. In relation to our ansatz (3) this side constraint has two advantages: First of all it is geometrically reasonable, since it ensures a bound on each coefficient of the polynomial. Second, it leads to a nicely solvable minimization problem. All in all curve recognition in our case is stated as

$$\min_{\|\beta\|_2=1} \|P\beta\|_2 \tag{4}$$

The above minimization problem leads to a nicely structured Eigenvector-Eigenvalue analysis as the following considerations show. We have

$$\min_{\|\beta\|_2=1} \|P\beta\|_2 = \min_{\|\beta\|_2=1} \sqrt{\beta^T P^T P \beta},$$

which is minimized by an eigenvector of $P^T P$ corresponding to an eigenvalue λ of minimal absolute value. The minimum itself is this very eigenvalue λ. In terms of a singular value decomposition of P, λ is no more than a singular value of P of minimal absolute value and (4) is minimized by a corresponding right singular vector of P.

Proof: $P^T P$ is symmetric and features an orthogonal diagonalization employing eigenvectors. Thus $\exists Q \in \mathcal{O}(k) : Q^T P^T P Q = \Lambda$ with a diagonal matrix $\Lambda = diag(\lambda_1, \lambda_2, \ldots, \lambda_k)$ exhibiting the eigenvalues λ_i $(1 \leq i \leq k)$ of $P^T P$. The j-th column of Q is the eigenvector corresponding to the j-th diagonal entry of Λ. Abbreviating $Q^T \beta$ by γ, we get:

$$\min_{\|\beta\|_2=1} \|P\beta\|_2 = \min_{\|\gamma\|_2=1} \sqrt{\gamma^T \Lambda \gamma}.$$

Let λ_j be an eigenvalue of minimal absolute value $|\lambda_j| = \min\{|\lambda_j| \mid 1 \leq j \leq k\}$ and let $e_j^T = (0, \ldots, 0, 1, 0, \ldots, 0)$ be the j-th canonical basis vector. Then

$$\min_{\|\gamma\|_2=1} \sqrt{\gamma^T \Lambda \gamma} = \sqrt{e_j^T \Lambda e_j} = \sqrt{\lambda_j}.$$

By undoing the transformation from β to $\gamma = e_j$, we get $\beta^T = (q_{1,j}, q_{2,j}, ..., q_{k,j})$, the j-th eigenvector of $P^T P$. If the minimal absolute eigenvalue is a single eigenvalue, then we get a unique eigenvector minimizing eigenvector β. If not, then β may be any vector in the span of all eigenvectors to the eigenvalues with an absolute value of λ_j. Since a singular value of P is the root of an eigenvalue of $P^T P$, the right singular vector corresponds to the eigenvectors from above. $\square$

In practice with large $\mathcal{P}$ the minimal absolute eigenvalue of $P^T P$ will always be unique. In cases, where $\mathcal{P}$ does not specify a singe curve and a hole bunch of curves could have $\mathcal{P}$ as sample point set, we can not expect an algorithm to always return the correct curve.

Let us assume having the sample points on a branch B of a yet unknown degree d_b in a no-numerical-noise environment. If we choose an arbitrary testing degree $d \in \mathbb{N} \backslash \{0\}$ and examine the behavior of P and (4), we have the following cases:

$d = d_b$: As we investigate in homogeneous polynomials, the length of the parameter vector β is $k_d = \frac{1}{2}(d+1)(d+2)$, but the corresponding curve is determined by $k_d - 1$ points in general position. Per assumption, $\mathcal{P}$ contains this many points in general position and thus $rank(P) = k_d - 1 = rank(P^T P)$ and $P^T P$ has a single vanishing eigenvalue $\lambda = 0$. The corresponding eigenvector is the desired parameter vector of the curve.

$d < d_b$: An analysis of the corresponding P yields: P is of maximal rank and $P^T P$ has no vanishing eigenvalue.

$d > d_b$: Let $d = d_b + r$ for some $r \in \mathbb{N}$. An analogous observation shows that $P^T P$ has $k_r = \frac{1}{2}(r+1)(r+2)$ vanishing eigenvalues. This is a reasonable behavior because it shows that the resulting curves are degenerate: One component is the curve of degree d_b fixed by $\mathcal{P}$ and the other component is a more or less random curve of degree r.

This shows that for a given $\mathcal{P}$ the product $P^T P$ has a vanishing eigenvalue $\lambda = 0$ if and only if the testing degree is greater than or equals to d_b. Thus for constructed curves, we can compute the eigenvalues λ of $P^T P$ of minimal absolute value for testing degrees $d = 1$, $d = 2$, ..., successively. This leads to a point where λ is zero. At that point we reached the desired degree.

In practice, where the sample points are disturbed by a small numerical noise, we will not observe vanishing eigenvalues but a significant drop (usually a few orders of magnitude) of the absolute minimal eigenvalues (see section 5). Thus we are able to say:

"You constructed a curve of degree d_b."

4   Type of Constructed Loci

Let $\mathcal{B}$ be the whole continuum of constructed branches B and let $d_\mathcal{B}$ denote the maximum of all degrees d_b of constructed curves B contained in $\mathcal{B}$. Since we deal with finite constructions, $d_\mathcal{B}$ is bounded. $\mathcal{P}$ is a discrete set of points originating from discrete positions of a semi-free mover of a construction. In the last section we have seen that $P^T P$ becoming (nearly) singular indicates that the degree of a curve is equal to or below a certain value. Due to the analyticity (omitting the details here[1]) almost any curve contained in $\mathcal{B}$ is of degree $d_{\mathcal{B}}$. Therefore we can select a generic parameter set in the parameter space of a specific construction and get a generic curve B. The degree d_b of B equals $d_{\mathcal{B}}$ with probability one. This means, we can detect $d_{\mathcal{B}}$ and prompt the user of a dynamic geometry program:

"You constructed a curve of degree d_b. Your construction will generically generate a curve of degree $d_{\mathcal{B}}$."

To further specify $\mathcal{B}$ with associated degree $d_{\mathcal{B}}$, we can calculate invariants with respect to either projective, affine or Euclidean transformations. If all curves that belong to a given construction show the same invariants stored in a database, a name like circles, conchoids or limaçon can be associated with $\mathcal{B}$.

One possibility to achieve this is to examine the orthogonal space $\mathcal{B}^\perp$. This is the space of all $k_{d_{\mathcal{B}}} = \frac{1}{2}(d_{\mathcal{B}} + 2)(d_{\mathcal{B}} + 1)$ dimensional vectors orthogonal to any coefficient vector of any curve in $\mathcal{B}$. Therefore $\mathcal{B}^\perp$ is an invariant of the underlying construction. Generating a matrix V containing at least $k_{d_{\mathcal{B}}}$ parameter vectors to generic curves of $\mathcal{B}$, $\mathcal{B}^\perp$ is the null space of V. It can be determined by an eigenvalue/eigenvector analysis: $\mathcal{B}^\perp$ is spanned by the eigenvectors of $V^T V$ to eigenvalues equal to zero.

The remaining task for a graduation of $\mathcal{B}$ would be to correlate the calculated orthogonal space with a database and retrieve a corresponding name. Then we can say something like:

"Your construction will generically generate a circle."

The idea of how to avoid costly calculations of $\mathcal{B}^\perp$ is to simply look up some orthogonal spaces for curve classes of degree $d_{\mathcal{B}}$ in a database and perform a multiplication with V. Getting a zero result we know $\mathcal{B}$ to be a subset of all curves with the orthogonal space taken.

As a simple example let $d_{\mathcal{B}} = 2$ and let $\mathcal{B}$ be a set of circles. We assume that all coefficient calculations are performed with the monomial basis for quadrics (see example after equation (3)). Looking in a database for orthogonal spaces to curve types of degree two, we will find *circles* with an associated orthogonal space $\mathcal{B}^\perp = span((1,0,0,-1,0,0),(0,1,0,0,0,0)) =: span(o_1, o_2)$, because circles are characterized by the following properties properties

1. the coefficient of x^2 equals the coefficient of y^2:
   $(1,0,0,-1,0,0) \cdot \tau_2(x,y,z) = x^2 - y^2 = 0, \forall(x,y,z)$ on a circle.
2. the coefficient of xy equals zero:
   $(0,1,0,0,0,0) \cdot \tau_2(x,y,z) = xy = 0, \forall(x,y,z)$ on a circle.

If in fact all members of $\mathcal{B}$ are circles, $V \cdot o_1$ and $V \cdot o_2$ are zero or almost zero due to numerical effects.

---

[1] See [1] for an in-depth analysis.

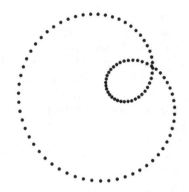

Fig. 2. Raw data of a degree-four-curve

To specify a more complex $\mathcal{B}$, a linear test like matrix multiplication may not suffice. It has to be tested wether potentially non-linear expressions hold for $k_{d_{\mathcal{B}}}$ generic curves out of $\mathcal{B}$. Additional difficulties arise when $\mathcal{B}$ is tested to be a subset of a parametrized class, like the class of conchoids[2]. Here, research is in progress to unify and simplify the tests of type affiliations. Focusing on Euclidean graduation, a promising approach seems to be to calculate an intrinsic rotation invariant center of a curve using all curve parameters as introduced in [7]. Then mapping $\mathcal{B}$ to a class of curves in a database by transformations is reduced to comparing invariants under rotations and scaling.

5   Experimental Results

Constructing a curve C in a dynamic geometry program and applying a curve recognition algorithm based on minimizing (4) yields some curve parameters. This corresponding curve should fit the sample data if the corresponding eigenvalue is significantly small. We present some data of numerical experiments with loci whose sample points have been calculated with the program Cinderella [4].

5.1   Finding the Degree

We start with the example of a limaçon corresponding to the first picture in Figure 1. The corresponding sample data $\mathcal{P}$ of the locus corresponds to the cloud of points given in Figure 2. The numerical precision of the data is approximately 14 digits. The Euclidean coordinates of points $p \in \mathcal{P}$ range from -10 to 10. Figure 3. shows a sequence of plots for the estimated curves of degree $d = 2, \ldots, 6$. The smallest absolute values of eigenvalues λ_d in these five situations are given below the pictures. The sample points are given for reference. One observes a significant drop of the eigenvalues from $d = 3$ to $d = 4$. The

---

[2] A conchoid symmetric to the x-axis and with the singularity in the Euclidean origin can be written as $b(x, y, z) = (x - \sigma)^2(x^2 + y^2) - \rho^2 x^2 = 0$ for some parameters σ, ρ.

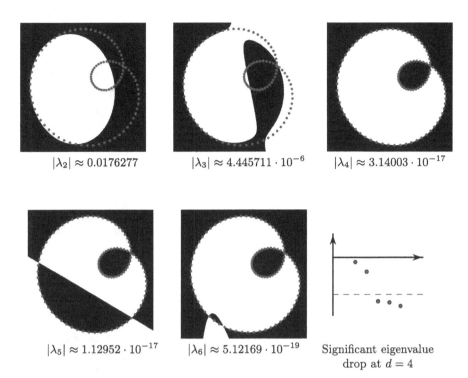

| $\|\lambda_2\| \approx 0.0176277$ | $\|\lambda_3\| \approx 4.445711 \cdot 10^{-6}$ | $\|\lambda_4\| \approx 3.14003 \cdot 10^{-17}$ |

| $\|\lambda_5\| \approx 1.12952 \cdot 10^{-17}$ | $\|\lambda_6\| \approx 5.12169 \cdot 10^{-19}$ | Significant eigenvalue drop at $d = 4$ |

Fig. 3. Estimated curves for degrees 2, 3, 4, 5 and 6 with logarithmic plot of the minimal eigenvalues against the tested degrees and threshold-line corresponding to $\lambda = 10^{-14}$

picture in the bottom right shows the logarithm of the minimal eigenvalues with a threshold-line at $\ln(\lambda) = \ln(10^{-14})$ marked vertically against the tested degree $d \in \{2, 3, 4, 5, 6\}$. Moreover for $d = 5$ we obtain (as expected) three absolute eigenvalues below $\lambda = 10^{-14}$ and for $d = 6$ we obtain six absolute eigenvalues below 10^{-14}. The next larger absolute eigenvalues are around 10^{-6}. This suggests that the curve under investigation is of degree $d = 4$. For the degree five approximation the plot unveils an additional component of degree one. And for degree six an additional component of degree two shows up.

As a second example we examine a parametrizable Epicycloid

$$f(t) = \begin{pmatrix} r \cdot \cos(k \cdot t) - s \cdot \cos(l \cdot t) \\ -r \cdot \sin(k \cdot t) - s \cdot \sin(l \cdot t) \end{pmatrix} . \tag{5}$$

If we choose $r = 2.4$, $s = 3$, $k = 2$ and $l = 5$ we get samples depending on the range and sampling rate for t. Three instances of sample point sets are shown in Figure 4. In the top row $t \in [0; 2\pi)$, in the middle row $t \in [0; 40\pi)$ and in the bottom row $t \in [0; 10\pi)$. In any case 72 equidistant values are used. Thus we have an ordered sequence of sample points given by our parametrization. In dynamic geometry programs ordered samples can be calculated even if

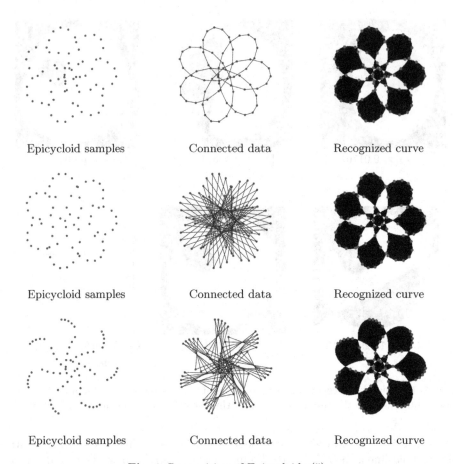

Epicycloid samples   Connected data   Recognized curve

Epicycloid samples   Connected data   Recognized curve

Epicycloid samples   Connected data   Recognized curve

Fig. 4. Recognition of Epicycloids (5)

the constructed curve is not rationally parametrizable[3]. A good idea of what the curve belonging to the data looks like can be provided by connecting the ordered samples linearly. The middle column of Figure 4 shows these line segments. In the case of $t \in [0; 40\pi)$ in the second row, the sampling rate is too low to show the curve itself but the picture gives a good impression of the contour. In case where the data is not scattered along the curve but locally concentrated, more intuition is needed when looking at the connected data. In either case our algorithm detects the correct curve of degree ten (ignoring roundoff errors).

As a further example, we take a Epicycloid of degree six with parameters $r = 1$, $s = 6$, $k = 12$ and $l = 4$. The top left image of Figure 5 shows it, correctly recognized by our algorithm. The corresponding minimal eigenvalue is sufficiently small, i.e. below the dashed threshold. The logarithmic plot of the minimal eigenvalues

---

[3] More precisely a curve can be parametrized by rational functions if it comes from a construction that uses exclusively ruler constructions. Constructions use ruler *and* compass may lead to more general (still algebraic curves).

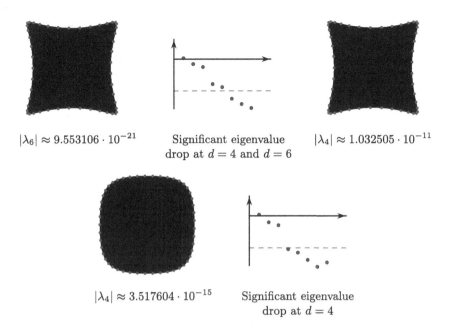

$|\lambda_6| \approx 9.553106 \cdot 10^{-21}$ Significant eigenvalue $|\lambda_4| \approx 1.032505 \cdot 10^{-11}$
                                                 drop at $d = 4$ and $d = 6$

$|\lambda_4| \approx 3.517604 \cdot 10^{-15}$ Significant eigenvalue
                                                 drop at $d = 4$

Fig. 5. Recognition of Epicycloids (5)

against the tested degree in the subsequent middle picture reveals another significant eigenvalue-drop. It suggests that the curve looks quite like a curve of degree
four. There is in fact such a curve of degree $d = 4$, approximating the sample
points very good. It can be seen in the rightmost picture in the top row of Figure 5.
We could have guessed this only by looking at the parameter vector of the correctly recognized curve: All parameters corresponding to the $x_1^i x_2^j x_3^k$-terms with
$i + j \in \{5, 6\}$ are very small or vanish totally (x_3 is the coordinate for homogeniza-
tion). This curve of degree four can be discarded because of a (second) significant
drop in the minimal eigenvalues when switching to a testing degree of six. An ac-
ceptable threshold, e.g. 10^{-14}, may be chosen to tell these curves apart. By altering
the Epicycloid's parameter r to $r = 0.4$, (5) still provides us with a Epicycloid of
degree six. In this case there is only one significant drop in the minimal eigenvalues.
λ_4 is already in the range of roundoff errors. Thus our presented algorithm falsely
assumes that the data belongs to a curve of degree four instead of six.

Our algorithm makes a false degree guess more often for curves of relatively
high degree. The may be recognized as curves of lower degree. Usually in most of
these cases the approximation by the low degree curve is so good that it visually
fits the sample data extremely well. The situation is qualitatively the same as
with the Epicycloid in Figure 5. For the example in the top row of the sample
data is already very nicely approximated by an algebraic curve of degree 4.

In general one can observe that the absolute value of the smaller eigenvalue
becomes smaller the higher the degree gets. Let us take for example the Epicycloid with a degree of 28 from Figure 6. It has the parameters $r = 1.4$, $s = 5$,

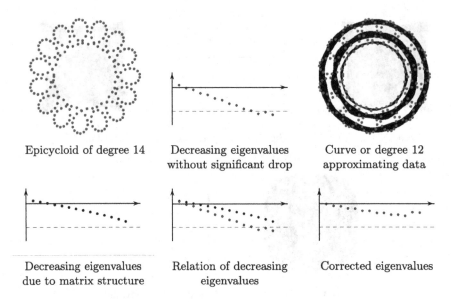

Epicycloid of degree 14	Decreasing eigenvalues without significant drop	Curve or degree 12 approximating data
Decreasing eigenvalues due to matrix structure	Relation of decreasing eigenvalues	Corrected eigenvalues

Fig. 6. Problematic recognition of an Epicycloid of degree 14 and structural decreasing eigenvalues

$k = 14$ and $l = 1$ and it comes up with a decreasing minimal eigenvalue the higher the selected tested degree is. This is not only the case with Epicycloids but with any curve. The minimal eigenvalues stop decreasing once they reach the region of roundoff errors. A reason for this behavior is in the structure of the matrix P. P is used to determine the degree in the minimization (4). It is built up rowwise by $\tau_d(p)$ with p being a sample point. Thus the columns of P consist of values $x^i y^j z^k$, where x, y and z are the coordinates of a sample point. Since the number of free parameters grows quadratically with the degree d curves of higher degree can simply approximates a set of sample points much better. We did not analyze this effect qualitatively but experimental results exploit a roughly exponential behavior. For this we we took a sample set of 300 random points within the range of -4 and 4 for x and y. With probability one these points will not be contained in a curve of degree 20 or lower. We can do this a hundred times and thereby calculate the mean minimal eigenvalues for our testing degrees. The resulting diminishing eigenvalues due to our matrix structure is shown in a logarithmic plot in the bottom left of Figure 6. In this logarithmic scale it is almost linear. The next picture (middle of lower row) compares this generic effect with the behavior for the Epicycloid. We see that at least part of the systematic eigenvalue fall can be explained by this generic effect. Knowing the structural diminution approximately, we can introduce a correction term in our calculations. The bottom right plot of Figure 6 reflects the outcome. This means that in case of our Epicycloid we have small eigenvalues for high testing degrees but not a single significant drop in the eigenvalues. The fact that λ_{12} in itself is lower than our chosen threshold is no indicator that a curve and

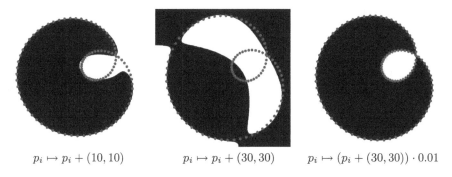

$$p_i \mapsto p_i + (10, 10) \qquad p_i \mapsto p_i + (30, 30) \qquad p_i \mapsto (p_i + (30, 30)) \cdot 0.01$$

Fig. 7. Experiments with shifted sample data for shifts $(10, 10)$, $(30, 30)$ and a shift $(30, 30)$ followed by scaling

the correct degree was found. If we would accept the curve at the testing degree $d = 12$, we would get the curve shown in the right of Figure 6. It is obviously not a good approximation to the sample points. In relation to the structural diminution, the minimal eigenvalues of the Epicycloid decrease even more. We interpret this as follows: The higher the testing degree gets, the better the data may be approximated. The increasing values in the corrected plot are due to roundoff errors.

5.2   Shifting and Preconditioning

Our method as presented here has one significant drawback: It is very sensitive to the location of the sample points. This is due to the following effect. If, for instance, in our example we shift the Euclidean sample points by a vector $(100, 100)$, then the corresponding parameters in $\tau_d(p_i)$ exhibit very high parameter values since the shifts are amplified by the large exponents in the polynomial bases. This results in the fact that our matrix P becomes more and more numerically ill-conditioned when the sample points are far away from the origin. The first two pictures of Figure 4 show how the estimated curve becomes more and more inaccurate the further away all sample points are form the origin.

So far we do not have a unified method to attack these numerical problems however we have several reasonable heuristics that work fine in practice. As a first step one could a priory investigate the data and translate it so that it is not too far away from the origin. One could also scale the data. After calculating the parameter vector this vector has to be transformed correctly to match the original data points again. We call this process preconditioning. The third picture of Figure 7 demonstrates the result of this method. There the data points have first been translated by a vector $(30, 30)$. This would normally result in a very badly conditioned matrix and the estimated curve would be far off the sample points (see middle picture). However now the whole data set is scaled by a factor of 0.01 which again moves the data points close to the origin. The last picture shows that after this preconditioning the estimated curve nicely matches the sample data again.

A more subtle method to attack numerical instabilities can be achieved by using topological curve invariants. In our example the middle picture could never stem from a locus generated by a dynamic geometry program: The curve obviously breaks up in (at least) two branches and $\mathcal{P}$ contains points on different branches. This is also true for the wrongly recognized Epicycloid of Figure 6. Using, for example, a Bernstein Basis (with respect to the appropriate bounded domain in each case) for representing and dealing with the involved polynomials could help to improve the conditioning problem, too.

5.3   Investigating Curve Invariants

The parameter vector of the curve is the eigenvector that corresponds to the smallest absolute eigenvalue. A search for curve invariants as proposed in Section 4 would require the generation of many instances of similar curves and the calculation of the orthogonal space. The orthogonal space would hint to specific dependencies on the curve's parameters. However, very often such dependencies also are indicated if one investigates only in one such parameter vector. In our example of Figure 2 after multiplying by a suitable factor the eigenvector of the degree four approximation takes the following form (four digits after the decimal point are shown):

$$\beta = (\ 1.0,\ -9.4034 \cdot 10^{-8},\ 2.0,\ -8.8127 \cdot 10^{-8},\ 1.0,\ -28.08,\ -0.55999,\ -28.08,$$
$$-0.55999,\ 237.3311,\ 7.8623,\ 40.2880,\ -423.9761,\ 66.5167,\ -1396.3979)$$

Looking at these values one may suspect that the curve has the characteristic properties

$$\beta_2 = \beta_4 = 0, \qquad 2\beta_1 = \beta_3 = 2\beta_5, \qquad \beta_6 = \beta_8, \qquad \beta_7 = \beta_9.$$

In fact, comparing these conjectures with the coefficients of a generic limaçon generated by a computer algebra system shows that the above relations indeed hold for the general case. Using techniques like the PSLQ algorithm [8] that is able to "guess" integral relations between real numbers one could also use a single eigenvector to derive many more reasonable conjectures about the underlying curve type. Further research in this direction is in progress.

6   Conclusions

With a given set of points $\mathcal{P}$ representing an algebraic curve approximately, we can determine the curve's coefficients by computing eigenvalues and eigenvectors only. This article presented the first steps along this road. Still there are many open questions and problems to attack. The main research problems currently are:

- What are good ways of preconditioning?
- How to deal with curve types that depend on parameters?
- How to derive geometric transformations that map a curve to a kind of standard representation?

- How can one derive a reliable measure for the quality of the result?
- To what extent can randomization techniques be used to speed up the calculations?
- How to use topological curve characteristics to restrict the search space of potential curves?
- Can similar approaches be used within the more special class of rational algebraic functions?

References

1. Kortenkamp, U., Richter-Gebert, J.: Grundlagen dynamischer Geometrie, Zeichnung - Figur - Zugfigur, 123-144 (2001)
2. Kortenkamp, U., Richter-Gebert, J.: Complexity issues in dynamic geometry. In: Rojas, M., Cucker, F. (eds.) Festschrift in the honor of Stephen Smale's 70th birthday, pp. 355–404. World Scientific, Singapore (2002)
3. Laborde, J.-M., Bellemain, F.: Cabri-Geometry II. In: Texas Instruments, pp. 1993–1998.
4. Richter-Gebert, J., Kortenkamp, U.: Cinderella - The interactive geometry software, Springer, see (1999), also http://www.cinderella.de
5. Taubin, G.: An improved algorithm for algebraic curve and surface fitting. In: Proc. ICCV 1993, Berlin, Germany, pp. 658–665 (May 1993)
6. Taubin, G.: Estimation of Planar Curves, Surfaces and Nonplanar Space Curves Defined by Implicit Equations with Applications to Edge and Range Image Segmentation. IEEE Transactions on Pattern Analysis and Machine Intelligence 13, 1115–1138 (1991)
7. Tarel, J.-P., Cooper, D.B.: The Complex Representation of Algebraic Curves and its Simple Exploitation for Pose Estimation and Invariant Recognition. IEEE Transactions on Pattern Analysis and Machine Intelligence 22, 663–674 (2000)
8. Ferguson, H.R.P., Bailey, D.H., Arno, S.: Analysis of PSLQ, An Integer Relation Finding Algorithm. Mathematics of Computation 68(225), 351–369 (1999)
9. Lei, Z., Tasdizen, T., Cooper, D.B.: PIMs and Invariant Parts for Shape Recognition. In: ICCV 1998. Proceedings of Sixth International Conference on Computer Vision, pp. 827–832 (January 1998)
10. Tasdizen, T., Tarel, J.-P., Cooper, D.B.: Algebraic Curves That Work Better. In: CVPR 1999. IEEE Conference on Computer Vision and Pattern Recognition, pp. 35–41 (June 1999)

Algorithmic Search for Flexibility Using Resultants of Polynomial Systems

Robert H. Lewis[1] and Evangelos A. Coutsias[2]

[1] Fordham University, New York, NY 10458, USA
[2] University of New Mexico, Albuquerque, NM 87131, USA

Abstract. This paper describes the recent convergence of four topics: polynomial systems, flexibility of three dimensional objects, computational chemistry, and computer algebra. We discuss a way to solve systems of polynomial equations with *resultants*. Using ideas of Bricard, we find a system of polynomial equations that models a configuration of quadrilaterals that is equivalent to some three dimensional structures. These structures are of interest in computational chemistry, as they represent molecules. We then describe an algorithm that examines the resultant and determines ways that the structure can be *flexible*.

1 Introduction

This project results from the recent convergence of four topics: systems of polynomial equations, flexibility of three dimensional objects, computational chemistry, and computer algebra.

Protein folding has been a major research topic in computational chemistry for a number of years [9]. Proteins are long molecular chains. Proteins form as flexible chains but they quickly fold into shapes that are rigidified by the formation of additional bonds. However, they retain flexibility in certain regions, which is essential for performing their various functions [25]. As macromolecules, composed of relatively heavy atoms (Carbon, Oxygen, Nitrogen, etc.) their conformational problem is modeled in terms of frameworks, i.e., systems of ideal points (atoms) connected by rigid rods (molecular bonds), with fixed angles (the bond angles, determined by molecular orbitals) but flexible torsions (the solid angles formed by successive bonded quartets) [13]. Simple examples are easily built using plastic balls and sticks.

In 1812, Cauchy considered flexibility of three dimensional polyhedra (think of a geodesic dome) where each joint can pivot or hinge. He proved that a convex polyhedron with invariant facets must be rigid [4]. Bricard [2], in response to a question posed in 1895 by C. Stephanos [24], gave the geometric conditions under which an octahedron may be flexible. The Bricard octahedra, however, besides being non-convex are also non-embeddable in 3-dimensional space as they possess intercrossing facets. A genuine, embeddable, flexible polyhedron with rigid facets was found by Connelly in 1978 [5], and soon models appeared of a simple flexible structure [8]. It is very enlightening to hold one of these and feel it move.

F. Botana and T. Recio (Eds.): ADG 2006, LNAI 4869, pp. 68–79, 2007.

As the facets are triangular, they are rigid – unlike quadrilaterals which are inherently flexible; think for example of a cube as compared to a tetrahedron. Thus the deformability is seen in terms of changes of the dihedrals formed by these facets about the edges of the polyhedron. Since the underlying description is in terms of quadratic distance constraints, expressing the conformational problem of the polyhedron in terms of cosines and sines (or half-tangents) of these dihedral angles results in systems of polynomial equations, quadratic in each of these variables, and in which the edge lengths enter as parameters. A polyhedron composed of triangular facets is subject to enough such constraints that according to classical results on rigidity that date back to Lagrange [12] and Maxwell [19] it should be rigid – generically at least! The polynomial system describing these conformations must therefore generically possess a discrete solution set. However, when conditions for flexibility are met, the solution set must acquire components of nonzero dimensionality (so called *continuous* components). Thus, the problem of detecting flexibility amounts to being able to identify conditions in the parameters for which a system of n polynomials in n variables drops in *rank* [23].

Here we present a new approach to understanding flexibility, using resultants and symbolic computation. The geometry of the object or molecule is described by a set of multivariate polynomial equations. Solving a system of multivariate polynomial equations is a classic, difficult problem. The approach via resultants was pioneered by Bezout [1], Dixon [10], and others. The resultant *res* appears as a factor of the determinant *det* of a matrix containing multivariate polynomials. But often *det* is too large to compute or factor, even though *res* is relatively small. We will describe a heuristic that overcomes the problem here, and in other cases [16]. Once we have the resultant, we describe an algorithm that examines the resultant and determines ways that the structure can be flexible. We discover in this way the conditions of flexibility for an arrangement of quadrilaterals in [2]. This system was posed by Bricard as an easily realizable, mathematically equivalent alternative to his flexible octahedra.

All computations below were done with Lewis's computer algebra system *Fermat* [14], which excels as polynomial and matrix computations [20]. We used a 1.8 ghz Macintosh G5, with 2 gigabytes of RAM, new in 2003.

2   Accelerating the Dixon Resultant

The Dixon Resultant method [10], following an idea of Bezout [1] and modified by Kapur et. al. [11], is presented in [11], [3], and [17]. Given a system of n polynomial equations

$$f_i(x_1, x_2, x_3, \ldots, a, b, \ldots) = 0, \ i = 1, \ldots, n$$

in $n-1$ variables x_i and a number of parameters $a, b, \ldots$, the method computes its *resultant*, i.e. a single polynomial in the parameters encapsulating the solution (common zero) to the system. A common variation is to have n equations in n

variables. Then one of them, say x_1, is considered a parameter to bring this into the previous form. In either case, the other variables have been *eliminated*.

In this paper all polynomials have coefficients in the ring of integers Z and the solutions are in the field C of complex numbers. All computations are exact.

The basic Bezout-Dixon idea is to construct a square matrix M whose determinant $det \neq 0$ is a multiple of the resultant. The factors of det that are not the resultant are called the *spurious factors*, and their product is sometimes called *the spurious factor*.

The naive way to proceed is to compute det, factor it, and separate the spurious factor from the actual resultant. Deciding what is spurious and what is the resultant is not always simple. However, when the original problem is based on geometry (as is the present problem) and one knows that the solution set is discrete, the resultant must involve all the parameters. (Otherwise, one parameter could have arbitrary values without affecting the variables. This will not occur in any realistic problem.) Typically, many factors of det do not involve all the parameters. Also, it is usually easy to simply plug in a known numerical solution and see which factor it satisfies. (The Dixon method is not guaranteed to work if the solution set is infinite; see [3]).

A graver problem is that the determinant may be so large as to be impractical or even impossible to compute, even though the resultant is relatively small; the spurious factor is huge. Further, the determinant may be so large that factoring it is impractical.

To overcome these problems, Lewis has developed several heuristic methods [16]. A method called EDF, Early Discovery of Factors, makes use of the existence of spurious factors. We reproduce it here for the convenience of the reader. By elementary row and column manipulations (Gaussian elimination), it discovers probable factors of det and plucks them out of $M_0 \equiv M$. Any denominators that form in the matrix are plucked out. This produces a smaller matrix M_1 still with polynomial entries, and a list of discovered numerators and denominators. Iterate. Here is a summary:

Algorithm EDF: Variation of Gaussian elimination to discover factors of the determinant.

Input: square matrix M. Let n = number of rows of M. All entries of M are polynomials. Assume $Det(M) \neq 0$.

Output: list of polynomials whose product is $Det(M)$.

```
Let num be a list of numerator polynomials, initially empty.
Let den be a list of denominator polynomials, initially empty.
Let M[i] be the submatrix of M from entry (i,i) down to (n,n).

for i = 1 to n do
  for j = i to n do
    Find the GCD of all entries in row j.
    Factor it out; append it to num.
    Find the GCD of all entries in column j.
```

```
      Factor it out; append it to num.
   endfor;
   Find a good pivot in M[i]. Move it into position (i,i).
   Do one step of Gaussian elimination using the pivot.
   for j = i to n do
      Find the LCM of all denominators in row j.
      Multiply it by row j; append it to den.
   endfor;
   if desired(i)
      { every fifth or tenth row is  reasonable. }
      Look for common factors in num and den lists.
      Consolidate num, den by dividing out such factors.
   endif
endfor.


Consolidate num and den lists. den should be empty.
Output num.
```

Notes:

- The resultant is usually in the numerator list. It is often the last entry. The remaining entries in the numerator list are then the spurious factors. Almost always the numerator list is long and interesting.
- If the determinant is irreducible, the final list of numerators must be trivial, i.e., just that one polynomial. But if it is not irreducible, there is no guarantee that the final list of numerators will be nontrivial.
- The "consolidate" step, in which we look for a common gcd among the numerator and denominator lists, can be scheduled in various ways, and this can have a noticeable affect on performance. There is no obviously best method. Experiments show that consolidation should be done every five to ten rows.
- The definition of "good pivot" is also not rigorous. Basically, one wants the "smallest" nonzero entry, so that the ensuing rational function arithmetic yields "small" entries in the rest of the matrix. Heuristics can be written depending on the number of terms, number of variables, their degree, and the size of the numerical coefficients.

EDF can work efficiently because *det* usually has many factors. This is a bad way to compute the determinant of a random matrix. But the Dixon matrices M are far from random. The total CPU time with this method is not always less than that of a standard determinant method; sometimes it is much more. We will see below that this technique can be dramatically successful. For other examples, see [16].

3   Flexibility of Polyhedra, and Computational Chemistry

This is a very old question. In two dimensions we may consider triangles, quadri- laterals, parallelograms, or more general n-gons. We imagine they are made of rigid rods connected by pins that are free to pivot at ideal joints. Triangles are obviously rigid, and any quadrilateral is flexible, though parallelograms are more flexible in that the angle at any two sides can take on any value. In three dimensions we likewise consider elementary chemical models, or polyhedra with triangular faces (like a geodesic dome).

In 1812 Cauchy [4] proved that convex polyhedra must be rigid. In 1897 Bricard [2] investigated nonconvex octahedra and found three ways they could be flexible. However, his examples are not embeddable in 3-space; they are self-intersecting. So the question was left unanswered, do there exist flexible polyhedra in 3-space? Surprisingly, in 1978 Robert Connelly [5] gave an example, with 18 triangular faces. Steffen [18] found a flexible polyhedron with only 14 triangular faces and 9 vertices. Maksimov [18] proved that Steffen's is the simplest possible flexible polyhedron composed of only triangles. See also [22]. It was later proved by Sabitov [21] that the volume of any such flexible polyhedron is invariant as it flexes.

Coutsias et. al. [6] [7] showed that Bricard's ideas have application to computational chemistry, and generalized them to solve the problem of Loop Closure, leading to a general algorithm for computing localized torsional deformations of molecular loops in proteins and nucleic acids. Bricard [2] states that the conformational problem of the octahedron is mathematically analogous to that of a system of articulated quadrilaterals. Such systems were important in the late 19th century, with applications to the transfer of force or motion in mechanical devices like sewing machines and automobiles, and today to robotic manipulators. While generically flexible systems where the number of variables exceeds the number of constraints are ubiquitous, here we are concerned with *non-generic* flexibility, where the number of variables and constraints are equal. Then flexibility is encountered only when certain conditions hold among the parameters of the system.

In particular, we shall consider the flexibility of the planar group of three quadrilaterals in figure 1. Corners A, B, C, D, F are freely hinged. AD, DC, CB, BA, GF, FE, HI are rigid rods. The joints at G, H, I, and E can pivot.

4   Algorithmic Approach

We want to write a program that will determine conditions for the geometric figure to be flexible. Our method:

- Label the sides $e, b, s_1, \ldots, s_9$ (see figure 1).
- With elementary analytic geometry find six equations relating the sides to the three angles *alpha, beta, gamma* at the base.
- Eliminate most of the variables; compute the resultant.
- Find a way to tell from the resultant when the figure is flexible.

4.1   The Equations

Finding the equations is elementary. The variables are ca, sa, cb, sb, cg, sg (sines and cosines of base angles). There are eleven parameters, $e, b, s_1, \ldots, s_9$.

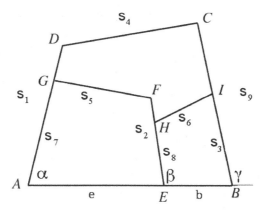

Fig. 1. Configuration of three quadrilaterals from Bricard [2]

Expressions for each x and y coordinate for each point $C, G, H, \ldots$ are easily found:

$cx := b + e + s_9 * cg;$
$cy := s_9 * sg;$
$gx := s_7 * ca;$
$gy := s_7 * sa;$
$hx := e + s_8 * cb;$
$hy := s_8 * sb;$
....

To form the six equations set each of these to 0 (the last three are just distances in the plane):

$sa^2 + ca^2 - 1,$
$sg^2 + cg^2 - 1,$
$sb^2 + cb^2 - 1,$
$(dx - cx)^2 + (dy - cy)^2 - s_4^2,$
$(ix - hx)^2 + (iy - hy)^2 - s_6^2,$
$(fx - gx)^2 + (fy - gy)^2 - s_5^2$

We have six equations, six variables, eleven parameters. The latter three equations are actually quite messy because the expressions $cx, cy, hx, hy, \ldots$ must be expanded in terms of the variables and parameters.

4.2 Solving the System with the EDF Method

We now apply the Dixon resultant method to the six equations, eliminating all variables but ca. The resultant will be a function of ca and the eleven parameters.

The Dixon matrix M is 29×29. The EDF method described in section 2 takes 62 minutes on the desktop Macintosh computer, and yields a list of numerators with more than 80 entries. The last two have 275808 and 312783 terms. Dividing each by their easily found contents c_1, c_2, yields the same polynomial of 201694 terms (recall that the *content* of a polynomial $f(x, \ldots)$ relative to x is the gcd of all the coefficients of $x^k, k = 0, \ldots, degree(f, x)$). This polynomial, call it *res*, is easily checked to be irreducible and is the resultant. It has degree 7 in *ca*. The product of the other terms in the list, call it h, has 35000 terms. The determinant of M, as the product of all these, $res^2\, h\, c_1\, c_2$, is truly gigantic, probably not computable on any computer. Thanks to EDF, there is no need to compute it.

4.3   Determining Flexibility

Classically, one would use the resultant by plugging in numerical values for the eleven parameters. That yields an equation in only the one variable *ca* that could be solved numerically. But how to detect flexibility in the quadrilaterals? Answer: If the parameters have the right relations to each other to produce flexibility, there are infinitely many values that work for *ca*. But *res* is a polynomial, so the only way to have that many roots is for every coefficient relative to *ca* in the resultant to vanish. That can be thought of as yielding eight new equations in the eleven sides, but those equations would be too complicated to use (their number of terms ranges from 198 to over 53000),

Instead, we have developed an algorithm *Solve* to produce a list of relations among the sides that will kill all eight coefficients. If the algorithm is good enough, any relationship producing flexibility will be on this list. However, it is not clear that all relationships on the list must produce flexibility. The issue of converse implication with the Dixon resultant is discussed in [3]. If a set of relations force all eight coefficients of *res* to vanish, when these relations are plugged into the original six equations, there is a positive-dimensional component to the solution set (i.e., a continuous family of solutions). But this may not be meaningful geometrically.

To describe the algorithm, let us first rename $e \equiv s_{10}, b \equiv s_{11}$. We present *Solve* in terms of general inputs f and x. f is a polynomial in x and N parameters s_i.

Algorithm *Solve(f, x)*: Given a polynomial f in a variable x and a number N of parameters s_i, find relations on the parameters that make the entire polynomial vanish. Our problem is solved by invoking $Solve(res, ca)$, $N = 11$.

Outline:

1. Kill each coefficient *coef* of x in turn, starting at the highest degree. Do so by looking for contents, linear parameters to solve for, or a difference of squares. When a substitution is found (it is possible that none will be found), plug it in, reducing the degree of f. Continue.
2. Whether or not substitutions were found in Step 1, also try to kill the coefficient *coef* by invoking the entire algorithm on it, relative to each variable in *coef*. So, this step of *Solve* works by calling $Solve(coef, s_i)$ within a loop.
3. Use suitable data structures to keep track of all the substitutions.

Here is a simple example. If res were $(s_9*s_8-s_7*s_6)ca^2+(s_4^2-s_3^2)ca+s_8-s_6$, one solution would be the collection (or *table*) of the three relations $s_9 = s_7, s_8 = s_6, s_4 = s_3$. On the other hand, the algorithm will fail on something random like $(5s_6^3s_9^6 + 2s_9^4 - 7s_4^3s_6^4s_9 + 1)ca^2 + (3s_5^4s_6^3s_9^7 + 2s_7^4 + 7)ca + s_5^4 + s_6^3s_9^5 - s_5^3s_6^4s_9 + 2$ as none of the techniques in Step 1 apply to the coefficients of ca.

The relations may be described as follows: Partition the set of N parameters into nonempty subsets $X = \{x_i\}_{i=1}^n$, $Y = \{y_j\}_{j=1}^m$, $n + m = N$. Each relation is an equation $y_j = g_j(x_{i_1}, x_{i_2}, \ldots)$ where g_j is a rational function. A collection of m of these for $j = 1, \ldots, m$ is a *solution table* if f evaluated at them all is 0. In the example above $X = \{s_3, s_6, s_7\}$ and $Y = \{s_4, s_8, s_9\}$.

Input: multivariate polynomial f in a primary variable x and N parameters s_i.

Output: list of solution tables, as defined above.

```
Let lst be the output list of solution tables, initially empty.
Let cc = leading coefficient in f(x).
{ cc is a polynomial in the parameters. }
Get factors of cc by finding content relative to all s_i.
{ Optionally, also do more complete factoring. }
Use the factors to produce a list ls of s_j to solve for:
    Within each factor, find all s_j of degree 1.
    Look for factors that are differences of squares.
{ Note: the list ls may be empty. }
while not done with the list ls do
    Solve for s_j, getting s_j = g(s_i_1, s_i_2, ...).
    Use the relation g to replace s_j in f.
    This yields fj(x), say, of lower degree.
    Compute lstj = Solve(fj,x).
    if lstj is not empty
        Insert the relation s_j = g into each table of lstj.
        Append the resulting lstj to lst.
    endif;
enddo;
for every s_i in cc not in ls do
    Compute lsti = Solve(cc, s_i).
    for every table T in lsti do
        plug the relations of T into f, yielding ft(x).
        Compute lstt = Solve(ft, x).
        Combine T with each table in lstt.
        Append the resulting lstt to lst.
    endfor;
endfor;

Look for duplicates in lst; "clean up" lst.
Output lst.
```

In creating the relations $s_j = g(s_{i_1}, s_{i_2}, \ldots)$, we reject any relation of the form $s_j = 0$, or in which all the numerical coefficients in g are negative, such as $s_3 = -s_2 - s_5 s_7$. Since the s_i are lengths on a geometric figure, these are meaningless.

Details of combining and managing the table lists are left to the programmer. Documented Fermat code for this algorithm is at [15].

There is no guarantee this method will work. However it does, in about 3 minutes. Thirteen tables are produced, all of which are variations or special cases of the following three. It finds the two ways to make the quadrilaterals flexible: all three are parallelograms, and one is a parallelogram and the other two are similar. Interestingly, it also finds a degenerate yet still meaningful arrangement when two of them are rhomboids.

For example, the case where the lower left quadrilateral is a parallelogram and the other two are similar is expressed by the table

$s_9 = s_3(e + b)/b,$
$s_8 = s1\ b/(e + b),$
$s_7 = s_2,$
$s_6 = s_4\ b/(e + b),$
$s_5 = e$

All three parallelograms is

$s_9 = s_1,$
$s_4 = e + b,$
$s_7 = s_2,$
$s_5 = e$
$s_6 = b,$
$s_8 = s_3$

Two rhomboids is

$s_9 = e + b,$
$s_8 = s_6,$
$s_4 = s_1,$
$s_3 = b$

Let us look more closely at the two rhomboids case. If those relations are substituted into the original six equations, one equation becomes extremely simple: $(b - s_6\ cb)(1 + cg) - s_6\ sg\ sb$. We are led to setting $cg = -1$ and $sg = 0$ (so $\gamma = \pi$), which kills three (not just two) of the six equations. Three remain, in the four variables sa, ca, sb, cb:

$sa^2 + ca^2 - 1,$
$sb^2 + cb^2 - 1,$
$2\ s_2\ s_7 sb\ sa + 2\ s_2\ s_7\ ca\ cb - 2\ e\ s_2\ cb + 2\ e\ s_7\ ca - s_7^2 + s_5^2 - s_2^2 - e^2$

We expect therefore a continuous family of solutions, which can be demonstrated by numerical experiments, or by computing the (bi-variate) resultant of the three equations, eliminating sa and sb:

$$8\,e\,s_2{}^2 s_7\,ca\,cb^2 - 4\,s_2{}^2 s_7{}^2 cb^2 - 4\,e^2 s_2{}^2 cb^2 - 8\,e\,s_2\,s_7{}^2 ca^2 cb + 4\,s_2\,s_7{}^3 ca\,cb -$$
$$4\,s_2\,s_5{}^2 s_7\,ca\,cb + 4\,s_2{}^3 s_7\,ca\,cb + 12\,e^2 s_2\,s_7\,ca\,cb - 4\,e\,s_2\,s_7{}^2 cb + 4\,e\,s_2\,s_5{}^2 cb - 4\,e\,s_2{}^3 cb -$$
$$4\,e^3 s_2\,cb - 4\,s_2{}^2 s_7{}^2 ca^2 - 4\,e^2 s_7{}^2 ca^2 + 4\,e\,s_7{}^3 ca - 4\,e\,s_5{}^2 s_7\,ca + 4\,e\,s_2{}^2 s_7\,ca + 4\,e^3 s_7\,ca -$$
$$s_7{}^4 + 2\,s_5{}^2 s_7{}^2 + 2\,s_2{}^2 s_7{}^2 - 2\,e^2 s_7{}^2 - s_5{}^4 + 2\,s_2{}^2 s_5{}^2 + 2\,e^2 s_5{}^2 - s_2{}^4 - 2\,e^2 s_2{}^2 - e^4.$$

The choice $cg = -1$, $sg = 0$ is actually geometrically meaningful. It corresponds to two degenerate rhomboids, with the points A, C and E, I falling on top of each other. Thus, the flexibility in this case is just the flexibility of the single quadrilateral $AEFG$. The original six equations do indeed fit this picture. See figure 2. If, on the other hand, we exclude the degeneracy of the angle γ, then it can be shown that the resulting problem has a resultant of lower degree, leading to additional, non-degenerate, discrete conformations. The identical vanishing of the resultant of the full problem for this case would completely mask the existence of these discrete components of the solution set.

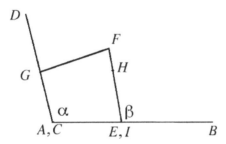

Fig. 2. Degenerate rhomboids

4.4 Future Work

By writing the equations in terms of the tangents of the half-angles, we can reduce the problem from six to three equations:

$$a_1 * t_1^2 * t_2^2 + b_1 * t_1^2 + 2c_1 * t_1 * t_2 + d_1 * t_2^2 + e_1 = 0,$$
$$a_2 * t_2^2 * t_3^2 + b_2 * t_2^2 + 2c_2 * t_2 * t_3 + d_2 * t_3^2 + e_2 = 0,$$
$$a_3 * t_1^2 * t_3^2 + b_3 * t_1^2 + 2c_3 * t_1 * t_3 + d_3 * t_3^2 + e_3 = 0$$

The t_i are the half-angle tangents of the three base angles. As before, these equations result from elementary analytic geometry. The parameters $a_i, b_i, \ldots$ are quadratic functions of the eleven sides. For example,

$$a_1 = e^2 + s_2^2 + s_7^2 - s_5^2 - 2e * s_2 + 2e * s_7 - 2s_2 * s_7$$

which is a product of two linear terms. This is the form of the equations as derived by Bricard.

The resultant of this system has 5685 terms. Shall we apply our flexibility searching algorithm as before? It is more subtle, as now we must try relations like $a_1 = 0$ or $a_1 = -d_3 - e_2$. When the parameters were actually the sides, substitutions like this made no sense and were excluded, thereby streamlining the search. We have recently modified algorithm *Solve* to consider these cases, and the work is ongoing. Success on this set of three equations would be significant because the identical set of equations arises in other contexts, and a variant (including also the "missing" terms, such as $t_1^2 t_2, t_3, \ldots$) gives the conformational equations of a protein or nucleic acid backbone [6], [7].

References

1. Bikker, P.: On Bezout's method for computing the resultant. RISC-Linz Report Series, Johannes Kepler University A-4040 Linz, Austria (1995)
2. Bricard, R.: Mémoire sur la théorie de l'octaèdre articulé. J. Math. Pures Appl. 3, 113–150 (1897), English translation:
   http://www.math.unm.edu/~{}vageli/papers/bricard.pdf
3. Buse, L., Elkadi, M., Mourrain, B.: Generalized resultants over unirational algebraic varieties. J. Symbolic Comp. 29, 515–526 (2000)
4. Cauchy, A.-L.: Deuxième mémoire sur les polygones et les polyèdres. J. de l'École Polyt. 16(1813), 87–99
5. Connelly, R.: A counterexample to the rigidity conjecture for polyhedra. Publ. Math. I. H. E. S. 47, 333–338 (1978)
6. Coutsias, E.A., Seok, C., Wester, M.J., Dill, K.A.: Resultants and loop closure. International Journal of Quantum Chemistry 106(1), 176–189 (2005)
7. Coutsias, E.A., Seok, C., Jacobson, M.J., Dill, K.A.: A Kinematic view of loop closure. Journal of Computational Chemistry 25(4), 510–528 (2004)
8. Cromwell, P.R.: Polyhedra, pp. 222–224. Cambridge University Press, New York (1997)
9. Dill, K. A.,,, Chan, H.S.: From Levinthal to pathways to funnels: The "new view" of protein folding kinetics. Nat. Struct. Biol. 4, 10–19 (1997)
10. Dixon, A.L.: The eliminant of three quantics in two independent variables. Proc. London Math. Society 6, 468–478 (1908)
11. Kapur, D., Saxena, T., Yang, L.: Algebraic and geometric reasoning using Dixon resultants. In: Proc. of the International Symposium on Symbolic and Algebraic Computation., A.C.M. Press, New York (1994)
12. Lagrange, J.-L.: Mécanique Analytique, Paris (1788)
13. Leach, A.: Molecular Modeling and Simulation, Cambridge (2004)
14. Lewis, R.H.: Computer algebra system Fermat. http://home.bway.net/lewis/
15. Lewis, R.H.: Fermat code for Solve. http://home.bway.net/lewis/
16. Lewis, R.H.: Heuristics to accelerate the Dixon resultant. to appear in Mathematics and Computers in Simulation
17. Lewis, R.,,, Stiller, P.: Solving the recognition problem for six lines using the Dixon resultant. Mathematics and Computers in Simulation 49, 203–219 (1999)
18. Maksimov, I.G.: Polyhedra with bendings and Riemann surfaces. Uspekhi Matemat. Nauk 50, 821–823 (1995)

19. Maxwell, J.C.: On the calculation of equilibrium and stiffness of frames. Phil. Mag. 27, 294–299 (1864)
20. Robertz, D., Gerdt, V.: Comparison of software systems (2004), http://home.bway.net/lewis/
21. Sabitov, I.: A proof of the "bellows" conjecture for polyhedra of low topological genus. Dokl. Acad. Nauk 358(6), 743–746 (1998)
22. Stachel, H.: Higher order flexibility of octahedra. Period. Math. Hung. 39, 225–240 (1999)
23. Sommese, A.J., Wampler II, C.W.: The Numerical Solution of Systems of Polynomial arising in Engineering and Science. World Scientific, New York (2005)
24. Stephanos, Cyparissos.: Problem 376. L' Intermédiaire des Mathématiciens, 2, 243–244 (1895)
25. Thorpe, M., Lei, M., Rader, A.J., Jacobs, D.J., Kuhn, L.: Protein flexibility and dynamics using constraint theory. Journal of Molecular Graphics and Modelling 19, 60–69 (2001)

Cylinders Through Five Points: Complex and Real Enumerative Geometry

Daniel Lichtblau

Wolfram Research, Inc.
100 Trade Center Dr.
Champaign, IL 61820 USA
danl@wolfram.com

Abstract. It is known that five points in $\mathbb{R}^3$ generically determine a finite number of cylinders containing those points. We discuss ways in which it can be shown that the generic (complex) number of solutions, with multiplicity, is six, of which an even number will be real valued and hence correspond to actual cylinders in $\mathbb{R}^3$. We partially classify the case of no real solutions in terms of the geometry of the five given points. We also investigate the special case where the five given points are coplanar, as it differs from the generic case for both complex and real valued solution cardinalities.

Keywords: Enumerative geometry, Gröbner bases, nonlinear systems.

1 Introduction

Given five generic points in $\mathbb{R}^3$ it is not hard to show that there are finitely many solutions to the set of equations that determine cylinders containing those points. This is to be expected because cylinders have five degrees of freedom (a radius and four parameters to determine the axial line). Several papers prove that the generic number of such solutions, in complex space, is six (counted with multiplicity) [4,7,8,13]. Of these any even number may be real valued.

This is of importance for several reasons. First we indicate two constraint geometry interpretations of the problem.

- Given five points in $\mathbb{R}^3$, find the smallest positive r and axis parameters such that the cylinder of radius $2r$ with those parameters tangentially encloses the balls of radius r centered at the points [8,22].
- A common need in scene classification [6] is to find a best fitting cylinder for a set of more than five data points. To do so one might start with an exact fit to five points, followed by optimization methods to get a best fit to the full ensemble. See [13] for several references to applications of this.

In order to tackle either of these it is necessary to find all cylinders through five given points [13]; clearly it is useful to know the expected size of the solution set (or an upper bound, if we restrict to only real valued solutions). In this paper we discuss several aspects to this enumeration problem.

F. Botana and T. Recio (Eds.): ADG 2006, LNAI 4869, pp. 80–97, 2007.

Some experimentation indicates that it is not infrequent, when working with pseudorandom point coordinates, to have configurations with no real cylinder solutions. We investigate this situation in an attempt to understand the geometry of the configurations that give rise to it. Another point of interest is the case where the five points are coplanar. In this situation we will see that the generic number of complex solutions to the cylinder parameter equations is four. It turns out, however, that the number of real solutions is at most two. In discussing the case of coplanar data we state a conjecture regarding the degenerate situation where there is a dimensional component to the cylinder solutions. This is of interest insofar as configurations near one with such a solution set might exhibit numerical instabilities in a computational setting.

The related problem of finding cylinders of a given radius through four given points in $\mathbb{R}^3$ is discussed in [9,10,15,22]. The special case of four coplanar points is discussed in [18]. In contrast to cylinders through five coplanar points, all cylinder solutions of given radius through four coplanar points can be real valued.

This paper is a companion to [13]. Substantial emphasis therein was placed on various aspects of solving systems and related computation for purposes of deducing properties of cylinders through five points. Here we rely primarily on elementary arguments that cover the theory, with less focus on computational specifics. Experiments that gave rise to this use code similar to that presented and explained in [13]. The rest of this paper is as follows. Section 2 proves that there are generically six (possibly complex valued) solutions to the equations for cylinders containing five given points. We begin by formulating polynomial conditions that the fourth and fifth points project to the same circle as the first three in a given direction, and work with the ideal formed by these polynomials. The main tactic is a count of solutions in projective space, followed by a simple computation to enumerate solutions at infinity. We then cover some implications of this. Section 3 discusses in detail the case of no real solutions. Here the main tool is again the cocircularity polynomials; we now make observations about the behavior of their respective direction curves in real space. In section 4 we look at cases that have six real solutions. Section 5 investigates the special case when the points are coplanar. This is followed with a summary and some directions of further inquiry.

One way to approach some of these problems, from the perspective of automated geometry, is to employ comprehensive Gröbner systems as in [20], to classify both generic solutions and specialized configurations e.g. where the number of solutions becomes infinite. The difficulty is that, to date, the computations involved have been intractable. One could regard the alternate approach taken in this paper as a blend of computational tools and human guidance, to make progress on problems for which fully automated methods seem to falter. That is to say, we do not work with full-blown automated geometric deduction tools, but borrow a bit from underlying computational methods.

In the sequel we use "cylinder" to denote solutions to polynomial equations for a cylinder, regardless of whether they are real or complex valued. To specify the former we use "real cylinders". Typically we will use the term "parameters" to refer to coordinates in the configuration space of the five points (which we may identify with $\mathbb{R}^{15}$ or even $\mathbb{C}^{15}$). The values that specify a cylinder, to wit, radius and axial line features, are generally referred to as "variables" (since they are what we solve for in finding cylinders) or as "cylinder parameters" to distinguish them from the coordinate parameters already mentioned.

I thank the anonymous referees of this and a prior draft. Their several useful remarks and suggestions improved both exposition and references.

2   Counting (Possibly Complex) Cylinders Through Five Points

We recall a way of setting up the problem that gives rise to two equations in two unknowns. This specific formulation is used in [13] but similar methods are given in [7,8]. We place one point at the origin, another at $(1,0,0)$, and a third in the xy coordinate plane at $(x_2,y_2,0)$. We project these onto the set of planes through the origin parametrized by normal vector $(a,b,1)$. In each such projection they uniquely determine a (possibly degenerate) circle. We obtain the two polynomials below by enforcing that the two remaining points, (x_3,y_3,z_3) and (x_4,y_4,z_4), project onto the same circle.

$$(-x_3y_2 - b^2x_3y_2 + x_3^2y_2 + b^2x_3^2y_2 + x_2y_3 + b^2x_2y_3 - x_2^2y_3 - b^2x_2^2y_3 +$$
$$2abx_2y_2y_3 - 2abx_3y_2y_3 - y_2^2y_3 - a^2y_2^2y_3 + y_2y_3^2 + a^2y_2y_3^2 - bx_2z_3 -$$
$$b^3x_2z_3 + bx_2^2z_3 + b^3x_2^2z_3 + ay_2z_3 + ab^2y_2z_3 - 2ab^2x_2y_2z_3 - 2ax_3y_2z_3 +$$
$$by_2^2z_3 + a^2by_2^2z_3 - 2by_2y_3z_3 + a^2y_2z_3^2 + b^2y_2z_3^2,$$
$$- x_4y_2 - b^2x_4y_2 + x_4^2y_2 + b^2x_4^2y_2 + x_2y_4 + b^2x_2y_4 - x_2^2y_4 - b^2x_2^2y_4 +$$
$$2abx_2y_2y_4 - 2abx_4y_2y_4 - y_2^2y_4 - a^2y_2^2y_4 + y_2y_4^2 + a^2y_2y_4^2 - bx_2z_4 -$$
$$b^3x_2z_4 + bx_2^2z_4 + b^3x_2^2z_4 + ay_2z_4 + ab^2y_2z_4 - 2ab^2x_2y_2z_4 - 2ax_4y_2z_4 +$$
$$by_2^2z_4 + a^2by_2^2z_4 - 2by_2y_4z_4 + a^2y_2z_4^2 + b^2y_2z_4^2) \quad (1)$$

One observes from this that the number of solutions is generically finite, and by Bezout's theorem it is moreover bounded by nine, as each polynomial has total degree of three in the variables (a,b). Moreover, using, say pseudorandom values for the coordinate parameters and solving for the cylinder parameters as per [8] or [13] one obtains six solutions. Thus we know there are generically at least that many solutions.

Before developing the theory it might be instructive to see how these curves intersect real space. We work with an explicit set of points: $(0, 0, 0)$, $(1, 0, 0)$, $(5/3, 3/4, 0)$, $(5/4,1, 4/5)$, $(3/4, 1/3, 1/2)$. Plugging these parameters into the polynomials above gives

$$(-9925 - 12960a + 9612a^2 + 2000b + 9000ab +$$
$$6480a^2b - 5713b^2 - 20160ab^2 + 12800b^3,$$
$$- 1063 - 324a + 144a^2 + 1014b + 792ab + 486a^2b -$$
$$559b^2 - 1512ab^2 + 960b^3) \quad (2)$$

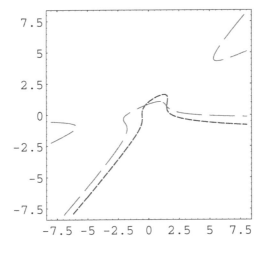

Fig. 1.

The zero sets are plotted in Figure 1. One sees from the curve intersections that there are two real solutions to the pair of equations, hence two real cylinders through this particular set of five points.

Proposition 1. *Configurations that give rise to an empty solution set or to a solution set of positive dimension lie on a variety in the configuration space.*

Proof (sketch). One obtains, in principle, a description of the generic solutions by forming a lexicographic Gröbner basis for the system (1). The process of doing this gives rise to the generic basis because at steps along the way one is allowed to divide by polynomials in the indeterminates. All inputs that fail to give the generic basis thus must satisfy conditions among the coordinates that cause these polynomials, upon specialization, to vanish. As there are finitely many steps in forming the basis, there are finitely many such conditions. As these vanishing conditions are defined by polynomials, their union is a variety. We may further refine it. Some configurations might fail to give the generic basis but still yield a nonzero finite solution set. If we exclude the conditions that give this situation, we are still left with a variety for which we get either zero or infinitely many (complex valued) solutions. □

Definition 1. *Above we saw that the subset of nongeneric configurations that give either zero or infinitely many solutions comprises a variety. We refer to this as the "bad variety", denoted V_{bad}. Several results below are stated in terms of configurations that miss this variety.*

We remark that V_{bad} is a part of the discriminant variety (see [12]).

Proposition 2. *There is a nonempty open set in our configuration space (which is, in effect, $\mathbb{R}^{15}$) for which we obtain no cylinders in $\mathbb{R}^3$.*

Proof. If one point is contained strictly within the convex hull of the other four then, as cylinders are convex, we have no real cylinders containing the five points. Small perturbations in real configuration space do not alter the situation that one is inside the hull of the other four. Hence there is an open set in real configuration space for which we obtain no real cylinders.   □

Corollary 1. *The maximum number of cylinders, already shown to be bounded by nine, is in fact no larger than eight.*

Proof. This is a consequence of the following facts. (i) Restricting to real inputs does not move us out of the generic case because this restriction is not algebraic. (ii) Given real data, complex solutions appear in pairs. (iii) The case where one point lies in the hull of the other four contains an open subset in the real part of the parameter space. Hence there is an open set in parameter space for which there are only complex solutions. So in general there must be an even number of solutions.   □

This also shows that the the number cannot be seven. So we know it is either six or eight.

Theorem 1. *Five generic points in $\mathbb{R}^3$ determine six distinct sets of cylinder parameters, of which an even number are real valued.*

Proofs may be found in [4,7,8,13]. They make use of various algebraic or geometric features particular to this problem. An algorithmic approach in [14] proves this blindly, that is, without use of geometry-specific features. (As with many algorithms in geometry, one can argue as to whether this is a good or bad thing, insofar as automated proofs often convey little insight. Regardless, algorithmic technology should not be ignored.) Proofs in [13] count roots based on either a Gröbner basis or resultant computation from (1). We give an independent proof below.

Proof. We will count the solutions at infinity for the polynomials shown in (1). We do this by homogenizing and setting the homogenizing variable to zero to get the initials (that is, the degree forms). They are
$$(-b^3 x_2 z_3 + b^3 x_2^2 z_3 + ab^2 y_2 z_3 - 2ab^2 x_2 y_2 z_3 + a^2 b y_2^2 z_3,$$
$$- b^3 x_2 z_4 + b^3 x_2^2 z_4 + ab^2 y_2 z_4 - 2ab^2 x_2 y_2 z_4 + a^2 b y_2^2 z_4)$$
The solution set for (a, b) consists of the three cases $a \to -b(1 - x_2)/y_2$, $a \to bx_2/y_2$, and $b \to 0$. We thus obtain three solutions at infinity for the homogenized system (these are simply the directions of the three lines between any pair of the first three points). The number of solutions from the Bezout theorem, nine, counts these three, and hence there are six solutions in affine space.   □

In [13] a mixed volume computation is done in order to bound the number of solutions using a method presented in [23]. The value obtained in this computation is 8, and from [23] we moreover know that that will be the generic number of solutions for problems with the same Newton polytope for monomial exponent vectors. To understand why this cylinder problem is not generic for the mixed volume computation, note that the initials are identical up to a constant multiple; any random perturbation (say, add b^3 to the first polynomial) will give only one solution at infinity, and therefore yield 8 finite solutions.

Proposition 3. *There is a closed set with nonempty interior in $\mathbb{R}^{15} - V_{bad}$ for which we obtain six real cylinders provided we count solutions by multiplicity.*

Proof (sketch). We may count the number of real roots using the Rule of Signs [19] on the univariate polynomial in any generic lexicographic Gröbner basis. This gives a closed condition for the boundary of the set of configurations that yield six sets of real valued cylinder parameters. To show it has nonempty interior it suffices to demonstrate one such configuration that has no multiple solutions. But this is the case for the two six-real-cylinder examples shown in [13] (one of which is formed from two regular tetrahedra sharing a common face).   □

Similar argument shows that the sets in $\mathbb{R}^{15} - V_{bad}$ that give rise to two or four cylinders real cylinders also have nonempty interiors. ¿From Proposition 2 we already knew this to be the case for configurations that give no real cylinders.

It may be useful to look at Theorem 1 in the context of what are known as comprehensive Gröbner bases [2,17,24]. This construction in effect allows one to circumvent the problem that Gröbner bases are not continuous in their input data; indeed it seems designed to address that defect. Such a basis contains encoded all Gröbner bases, for a given ideal with respect to a specified term order, under all specializations of the parameters. It does so in essence by doing multiple polynomial reductions on a given polynomial in the basis, to simultaneously allow for the possibility that any nonnumeric leading coefficient might or might not be zero. The upshot is that the coefficients of the comprehensive Gröbner basis vary continuously in the parameters of the configuration. The typical use of such a basis is in concrete examples when one wishes to make case distinctions based on parameter values. When a lexicographic term ordering is utilized we can say a bit about the structure of such bases in the (generic) case of finite solution sets, using insight gained from our examples.

For instance, suppose we have six distinct solutions in a situation where the Shape Lemma [1] does not apply (see [13] for two such examples involving six distinct real valued cylinders). We consider the basis for such a numeric specialization of the general problem. From Gröbner basis theory we know it contains a univariate polynomial in the lexicographically last variable (say, a). The degree of this polynomial must be less than six as we have assumed the Shape Lemma does not hold for this particular ideal and term ordering. We refer to the remaining variables as "higher".

We return for the moment to the general case wherein coefficients are again indeterminate parameters that vary in our configuration space. Note that a comprehensive Gröbner basis encodes, in vanishing conditions of leading coefficients, the basis for specializations of the sort just described. It also encodes the generic basis. By the Shape Lemma this latter contains linear polynomials in each of the higher variables with lead coefficients that generically do not vanish. By [11] all parametrized coefficients of at least one such linear polynomial must become identically zero for the nongeneric type of specialization under consideration (otherwise we would have fewer than six solutions). Hence the comprehensive basis must also contain nonlinear polynomials in those variables.

Now consider a generic specialization of the configuration parameters. Again invoking results from [11] we know that solutions are obtained from the subset of the comprehensive basis that encodes the generic case (that is, the generic degree six univariate polynomial and the polynomials that are linear in each of the higher variables). These are shown also to satisfy those polynomials that are nonlinear in the higher variables. So such a linear polynomial (say, in the variable b) must have the form $f(params)(b - g(a))$ where the second factor divides any corresponding polynomial(s) of higher degree in b. In other words, when the first factor, which involves only parameters, is nonzero, then where the second is satisfied all those in higher degree must also be satisfied; moreover they cannot all vanish when the first factor vanishes, so they must be divisible by the second factor.

3   Configurations with No Real Solutions

Theorem 2. *Suppose we have four noncoplanar points in $\mathbb{R}^3$. They are the vertices of some tetrahedron. Then there is an open set S containing the open tetrahedron and a dense subset of its boundary in the configuration space, such that if the fifth is chosen in S there will be no real cylinders containing all five points.*

Remark 1. If the fifth point is inside the convex hull of the other four then we already know this result. Now take the tetrahedron formed by the four points. Through each vertex the planes containing the three coincident faces form a cone with triangular base. If the fifth point lies within that cone then it obscures that vertex, i.e. the vertex lies inside the new tetrahedron formed by the fifth point and the remaining three. Hence this case is also covered by the "one point in the hull of the others" situation. Note that in this case the fifth point need not be near in distance to the other four.

Proof (1). Suppose the fifth point lies on a face of the tetrahedron formed by the other four. Then the convexity argument still tells us that no cylinder in $\mathbb{R}^3$ can contain all five points. As we assumed the tetrahedron coordinates are generic, we are in one of two situations: either having the fifth point lie on a face formed by three others puts the configuration in V_{bad} or it does not. We show that generically it does not, or in other words, the algebraic condition that four points are coplanar is not a condition for the bad variety. That this is so follows from the trivial observation (verified computationally) that there are configurations with four coplanar points that give rise to lexicographic Gröbner basis with the generic "shape"; were this a condition to lie in the bad variety then every configuration with four coplanar points would be in it.

The preceding argument shows that generically the fifth point is not on a tetrahedral face of the other four, so the set of such fifth points is dense in the set of all boundary points of the tetrahedron. By genericity we may assume that we have a univariate polynomial of degree six for one of the cylinder parameters. As there are no real cylinders containing this configuration, this polynomial has exclusively nonreal roots. These roots vary continuously with the configuration, hence the imaginary parts remain nonzero under small perturbations of the five points. Thus there is an open set around this point on the boundary for which we still obtain no real roots. As we require real roots in order to obtain cylinders in $\mathbb{R}^3$ this suffices to finish the proof.   □

Proof (2). This line of reasoning was suggested by [Richard Bishop, private communication]. In order for all five points to lie on a cylinder there must be a plane tangent to the unit sphere, and a circle in that plane, such that they all project onto that circle. Suppose the fifth point is inside or on the boundary of the tetrahedron formed by the other four. Then it is clear that the projection of the five points onto any such plane will have the projection of this last point contained in the quadrilateral formed by the projection of the other four. Hence any quadratic in the plane that contains all five projected points must be a hyperbola (because all other quadratic curves are convex). Moreover the parameters of the hyperbola equation are continuous in the locations of the five points. As the set of projection planes is compact, a small perturbation of the fifth point beyond the hull of the other four will not alter the situation that the five points project onto hyperbolas in all such planes, hence they cannot lie on any cylinder in $\mathbb{R}^3$. Hence from every boundary point on the tetrahedral hull of the four points, we may perturb outward some minimal distance (depending on that boundary point) and still have no real cylinders. As the tetrahedron boundary is compact we deduce that there is a minimum positive distance we can move outside and still not get real cylinders.   □

Corollary 2 (to proof). *There is an open set S containing the closed tetrahedron, such that if the fifth is chosen in S there will be no cylinders in $\mathbb{R}^3$ containing all five points. In other words, the "bad" variety in configuration space is not an issue.*

We now wish to show that all configurations that give no real cylinders arise in the setting of Theorem 2. Specifically we state the following conjecture.

Conjecture 1. Suppose we have a configuration of five points in $\mathbb{R}^3$ for which no real cylinders exist, and moreover assume that no point lies in the hull of the others. Then one of the points can be moved anywhere inside the convex hull of the full set and still we will get no real cylinders. In particular we could move this point along a line segment from outside to inside the hull of the other four, and at no point on that path would we get real cylinders. Thus we could regard the given configuration as a perturbation of one that has one point inside the hull of the other four, effectively providing a converse to Theorem 2.

We make observations of sufficient conditions for a proof and then state as a theorem a special case wherein we can fulfill the conditions. First consider the two curves that are solution sets to the two polynomials in (1). They are cubics that have one or more closed topological components in two dimensional real projective space, and because they are cubic they each have at least one real component that goes to infinity. (It is well known that they are connected as complex curves; by "components" we mean the obvious thing with respect to intersections with real space.)

We regard each point on such a curve as a solution to the direction parameter equations given by the four points in configuration space lying on a cylinder with axis in that direction. In other words, each point on the solution curve in parameter space defines a cylinder through the four points in configuration space that were used to form that equation. Suppose that at a solution on one such component, the fifth point lies inside the cylinder thus obtained (we are being loose with terminology but trust the meaning of "inside a cylinder" is clear).

Claim: For generic configurations it must then lie inside all cylinders defined by points on that affine component of the solution curve.

The proof of this claim has a small complication. Specifically, we must show that in order for the fifth point to "escape" outside the cylinder containing the other four, it must cross that cylinder (in contradiction to our hypothesis that there are no real cylinders containing all five points). A priori there is another way it might escape: the cylinder containing four can degenerate to a plane and subsequently reverse its "open" side. This degeneration can arise if the four points project onto a line for some direction. Such a direction must then lie in all four planes containing three of the four points. This gives an overdetermined and generically inconsistent set of linear conditions. We thus have verified the claim for generic configurations: if the fifth point is inside a cylinder containing the other four, then it stays inside the cylinders defined by all points on that component of the solution curve, and these cylinders each contain the other four points.

Next we observe that, were this true not just on one topological component of the direction solution curve, but on all of them, then that fifth point works in the conjecture. We see this as follows. When the fifth point lies inside all cylinders defined by a solution curve component, then it projects along the cylinder axis to a point inside the circle that intersects the projections of the other four. The same must hold for any other point in the interior of the convex hull of the five points. This is because such a point, written as a convex combination of the five, must either be in the interior of the tetrahedron defined by the first four (and thus project to the interior of the circle they define), or else have a nontrivial component of that fifth point and again thereby project to that circle interior.

We now proceed to construct a solution on one direction curve, that is, a cylinder containing four of the points, such that the fifth is inside it. We can arrange our four so that three are in the xy coordinate plane and the remaining two have a segment joining them that intersects the triangle defined by the first three (using a rotation, such an arrangement can always be found for a configuration of five points in $\mathbb{R}^3$). We place the fourth point on the z axis beneath the origin. Projecting from the fourth point onto a plane in the direction of the segment between the fourth and fifth points gives a unique circle containing the first three. The cylinder along that direction and containing that circle thus encloses the fourth and fifth points. Now we simply move one of the direction coordinates, forming new projections and cylinders containing the first three points, until one of the remaining points (say, the fourth) hits that cylinder. What we have done is to arrive on one of the two direction solution curves we set up in (1). Thus we obtain a cylinder containing four points and enclosing the fifth. From the discussion above we know that this holds for all cylinders defined by this affine component of the curve of directions.

At this point we have a sufficient condition for the conjecture to hold. We simply require that each of the solution curves have only one affine component.

Theorem 3. *Given five points in $\mathbb{R}^3$ for which there are no real solutions to the cylinder equations, suppose there are three such that*

(i) The segment joining the remaining two intersects the triangle bounded by those three.
(ii) The two curves of solution directions for cylinders containing those three and either the fourth or fifth respectively, each have only one component in real projective space.

Then either the fourth or fifth point can be moved anywhere inside the hull of the five and there will be no real cylinder containing this new point and the other four.

As remarked above we can always order the points in such a way that the first condition holds. But then in general the second condition will not hold. We believe the conjecture to be true all the same, though we do not have a proof at this time. We also mention that extensive graphical evidence suggests that most often these curves have one component in two dimensional real projective space. This is found by taking random examples with three points in the xy coordinate plane and the fourth and fifth above and below respectively, throwing away those that have real solutions, throwing away from the rest those for which the segment between fourth and fifth points does not go through the triangle bounded by the first three, and plotting the zero level sets for the two cylinder equations in remaining cases. "Most often", in this setting, refers to specifics of how we select our five points; we use pseudorandom values in a unit interval for each of the free parameters. In any case it would seem that this method applies frequently to configurations that give rise to no real cylinders.

We illustrate with a configuration that meets the conditions of the hypotheses. There are no real cylinders containing the set of points (0,0,0), (2,0,0), (1,2,0), (5/4,1,1/2), and (3/4,1,-1/3). This point set is shown in Figure 2. It may be seen that the segment joining highest and lowest points pierces the triangle formed by the other three.

Here are the direction parameter polynomials obtained by requiring that the fourth and fifth points respectively project onto the circles determined by the first three points and a given direction vector.

$$(-575 - 40a - 384a^2 - 40b - 200ab + 160a^2b - 159b^2 - 40b^3,$$
$$-207 - 24a - 128a^2 + 24b + 72ab - 96a^2b - 47b^2 + 24b^3)$$

Fig. 2.

Figure 3 shows plots of their respective zero sets. It is clear from the way the respective affine parts will meet at infinity that each has one topological component in real space, and moreover they do not intersect at finite points. Hence Theorem 3 applies to this configuration.

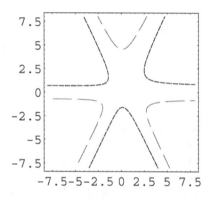

Fig. 3.

Note that we can weaken the second hypothesis of Theorem 3 so that the curves may have multiple components, provided the components for one are not separated by any component of the other. Graphical evidence supports the belief that this weaker requirement is always satisfied. Clearly a proof to this effect would suffice to prove the conjecture.

4   Configurations That Have Six Real Solutions

We describe in brief two cases that have six solutions. Further detail may be found in [13].

(1). Start with four points forming vertices of a square in the xy plane. This is the base of a pyramid with the fifth point as its apex above the centroid of this square. We obtain two horizontal cylinders each passing through a pair of opposite triangular faces of the pyramid. The remaining four each pass through a triangular face, angled upward, and an edge of the base. This case was first posted in [21].

(2). Take as five points the vertices of two regular tetrahedron sharing a common face. This presents an interesting set of symmeries. Any cylinder will have a "mirror image" obtained by reflection across the joined face, and also have two "conjugate" obtained by rotation of $\pi/3$. We show a visualization of this in Figure 4; the common face lies in the horizontal plane.

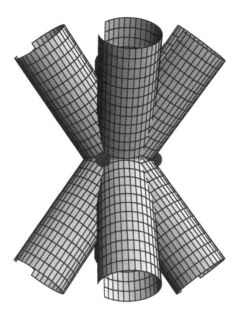

Fig. 4.

We now show the planar curves in real space that one obtains from the pair of cocircularity conditions in (1). We also perturb the five points slightly and show the resulting intersections. This is because the unperturbed case is a degenerate setup and the actual curves are each three lines. The perturbation indicates how each triad of lines can split at intersections into pairs of curves in the projective plane.

This configuration has some interesting properties. If the tetrahedron edge length is $\sqrt{3}$ then the common cylinder radius is $9/10$. It is related to a configuration from [15], wherein we have four points and a fixed radius for which there are twelve real cylinders through the points. For that we take vertices of one of the tetrahedra as our four points, and $9/10$ as the common radius. We obtain four sets of six cylinders by gluing tetrahedra respectively to each face of the given one. But these pair off for a total of twelve cylinders.

This configuration is also, perhaps surprisingly, related to the case of no real cylinders. We start with the doubled regular tetrahedra. There are two vertices not on the common face. Now let one of them move along the axis connecting it to the other. The two and three fold symmetry considerations indicated above imply that we either have six real cylinders (counting multiplicity), or none. It

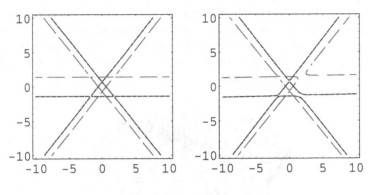

Fig. 5.

is not hard to realize that either moving it "too close" or "too far" from the opposite vertex gives rise to a configuration approaching the case of one point inside the hull of the other four. This is quantified in an explicit computation in [13]. It is also makes for an interesting dynamic geometry visualization to change the one vertex and depict the six cylinders coalescing into three pairs (multiple roots) before vanishing from real space [5].

It is an interesting question whether all cases of six real cylinders arise as "perturbations" of the two cases discussed above. As a starting point it would be useful to know whether they can be perturbed into one another with six real cylinders for every configuration along the perturbation.

5 Nongeneric Configurations

Thus far we have discussed exclusively the generic situation. It is of interest to make a few observations about the nongeneric case. This in turn sheds light on cylinder solutions for point configurations that are generic but "near" to such nongeneric ones.

Proposition 4. *Sets of five coplanar points are not generic insofar as they do not give six solutions to the cylinder equations. In general they give four such solutions.*

Proof. This follows from a straightforward computation with the pair of polynomials from (1). We substitute zero for the two nontrivial z parameters and compute a Gröbner basis in terms of the cylinder direction variables (a, b). This is in the form specified by the Shape Lemma and has a univariate polynomial of degree four with a second polynomial linear in the remaining variable. Hence for coplanar configurations there are generically four solutions rather than six. Of course there are further degeneracies that can arise. If, for example, four of the points are collinear then there will be infinitely many cylinders containing all five.

To finish the proof we must show that there are no cylinders parallel to the xy coordinate plane (as we tacitly set the z coordinate of the normal vector to 1). But this is clear from the fact that such a cylinder would intersect the coordinate plane in a pair of lines, and generically the five coplanar points do not lie on any pair of lines.   □

Corollary 3. *As configurations of five points move toward a generic coplanar configuration, two of the six (possibly complex) cylinder solutions go to infinity.*

This shows that in any comprehensive Gröbner basis for the system, using a lexicographic ordering, the univariate polynomial of generic degree six has leading and second coefficients vanish when the points are coplanar. The third coefficient will in general not vanish in this situation.

We now describe what happens in the generic coplanar case.

Theorem 4. *Given five coplanar but otherwise generic points in $\mathbb{R}^3$ there are four (complex) cylinders containing them. Of those, either zero or two will be real cylinders.*

Proof. That there are four complex cylinders was noted in the proof to Proposition 4. The five points uniquely determine a quadratic curve in the plane in which they lie, and generically it is either a hyperbola or an ellipse. The intersection of a cylinder with a plane is likewise a quadratic in that plane. Thus any cylinder containing five coplanar points contains the entire quadratic curve they determine. If that curve is a hyperbola then no real cylinder can contain it. If instead it is an ellipse then there are two real cylinders that contain it. These two cylinders have radial axes that each go through the center of the ellipse and lie in the plane perpendicular to the ellipse minor axis, and their angle of intersection is determined by the eccentricity of the ellipse.   □

For illustration we show in Figure 6 the case of two real cylinders when the points (all in the xy coordinate plane) are (1, 0, 0), (-1/3, 1, 0), (4, -1, 0), (1/2, 2/3, 0), and (1/4, -1, 0).

We can use the computational construction (1) to shed light on the problem of counting the number of cylinders of a given fixed radius through four points (which, as noted in [15], is equivalent to the problem of counting the number of lines simultaneously tangent to four given spheres of equal radius). As the radius is fixed (say, to 1), we are no longer free to rescale so we would use $(x_1, 0, 0)$ for our second point. We would project to circles using only two points along with the given radius. An important difference to arise is that for projections of two points onto any given plane, there are two circles of the given radius containing them. This ansatz would lead us to expect twice as many solutions for this problem as we obtained for counting cylinders through five points. That there are in fact twelve (not necessarily real valued) cylinders of given radius through four generically placed points is a theorem in [15]. In the special case that the points are coplanar, that there are eight such cylinders is a result of [18]. All of them can be real valued which is an interesting contrast to the result

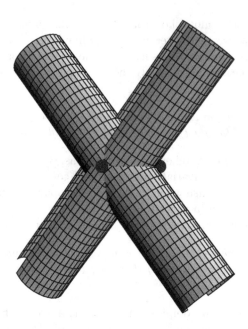

Fig. 6.

of Theorem 4 above. We remark that the generic number of complex solutions to these systems is obtained algorithmically in [14].

We would like a converse to Proposition 4. We begin with the observation that any configuration with four (or five) collinear points, or three collinear points and the remaining two on a parallel line, will contain infinitely many cylinders. We would like to know whether these are the only configurations for which that is the case. We note that a similar situation was shown in [3] for having infinitely many lines tangent to four given spheres.

Once three noncollinear points are fixed in a plane, there are finitely many ways to combine the remaining two points such that one of the above conditions holds. For any such combination, there are two degrees of freedom in how the points are placed. We will show that this is consistent with dimensional considerations.

Suppose we have infinitely many solutions. If they arise from but finitely many axial directions, then one can readily show that any infinite solution set comes about from four collinear points or points lying on two parallel lines, so that the axis is uniquely determined. We now assume this is not the case, that is, we have infinitely many axial directions. Consider our axial direction cubics from (1). In order to have infinitely many solutions, the algebraic curves that are solution sets to these two polynomials must share a component. That is, on a component of directions in which four points project onto a circle, the fifth must project onto that same circle.

Observe that cubic equations of the form imposed by our choices above have eight degrees of freedom (general cubics have ten coefficients but ours lose one degree three term, and the cubics are only defined up to nonzero scalar multiplication, giving eight degrees of freedom). Hence pairs of cubics of that form have sixteen parameters. The set of pairs we can actually attain has eight degrees of freedom (from the eight coordinates not a priori known). In order that a pair share a component, they must factor (they cannot be identical unless a pair of points coincides, in contradiction of hypotheses). The set of pairs that share a common factor has dimension ten. Thus we expect the dimension of the set of attainable cubic pairs that share a component to be given by the intersection dimension, which is two.

Conjecture 2. Any configuration of five distinct points for which there is a dimensional component to the cylinder parameter solution set must be coplanar. Either four points are collinear, or three are collinear with the remaining pair on a line parallel to that containing the first three.

This conjecture, alas, is not true as stated. In particular it does not hold in complex space. Here is a specific configuration violating the hypotheses, for which the solution set is infinite.

$$(0,0,0), (1,0,0), (-2, 8/5 - (6i)/5, 0), (2, -2, 1), (-2, 14/5 + (12i)/5, -3)$$

It remains an open question whether there are configurations with real coordinates which comprise a counterexample to the conjecture. Attempts to prove this via computational tactics (formulate relations based on factoring of the polynomials and equating a pair, finding relations among the data parameters, then checking that no real solutions can exist) have foundered to date due to computational complexity. That complex solutions can exist is already a bad sign insofar as it means one must use real solving technology e.g. cylindrical algebraic decomposition, and this is known to be computationally intensive.

6   Summary

We reviewed in brief the fact that there are generically six solutions to the equations for cylinders through five points in $\mathbb{R}^3$, noting that of these any even number, counting multiplicity, may be real valued. We discussed in some detail the case where there are no real valued solutions. Specifically we have a theorem and conjectured converse relating the case of no real cylinder solutions to perturbations of having one point within the convex hull of the other four. We also described configurations for which there are six real solutions.

We proceeded to the special case where the points are coplanar. In this situation generically there are only four solutions, and, in contrast to the noncoplanar case, at least two are not real valued. When two are real valued and distinct, small perturbations from coplanarity will not alter this situation. A natural question is whether there is a plausible conjecture, similar to that for the case of no real cylinders, to the effect that all cases of two real cylinders are perturbations of coplanar cases.

It would be interesting to get a geometrical description of what configurations will give rise to the other numbers of real valued cylinders. In [13] it is observed that all computational examples observed having six real solutions appear to be perturbations of the two particular configurations we mentioned. But this is quite far from a systematic understanding of the geometry of five point configurations that give six real solutions. We seem to know even less about the case of four real solutions.

Related to the geometric classification of where the number of real or complex solutions changes, there is the algebraic description via real and complex algebraic sets. From this point of view it would be nice to better understand the discriminant variety [12,16]. Unfortunately this seems to be computationally intractable using existing technology.

References

1. Becker, E., Marinari, M., Mora, T., Traverso, C.: The shape of the Shape Lemma. In: ISSAC 1994. Proceedings of the 1994 International Symposium on Symbolic and Algebraic Computation, pp. 129–133. ACM Press, New York (1994)
2. Becker, T., Kredel, H., Weispfenning, V.: Gröbner bases: a computational approach to commutative algebra. Springer, Heidelberg (1993)
3. Borcea, C., Goaoc, X., Lazard, S., Petitjean, S.: Common tangents to spheres in $\mathbb{R}^3$. Discrete & Computational Geometry 35(2), 287–300 (2006)
4. Bottema, O., Veldkamp, G.: On the lines in space with equal distances to n given points. Geometriae Dedicata 6, 121–129 (1977)
5. Bryant, J., Lichtblau, D.: Cylindersthroughfivepoints (2007), http://demonstrations.wolfram.com/CylindersThroughFivePoints/
6. Chaperon, T., Goulette, F.: Extracting cylinders in full 3-d data using a random sampling method and the Gaussian image. In: VMV 2001. Proceedings of the 6th International Fall Workshop Vision, Modeling, and Visualization, Stuttgart, Germany, Aka GmbH, pp. 35–42 (November 2001)
7. Chaperon, T., Goulette, F.: A note on the construction of right circular cylinders through five 3d points. Technical report, Centre de Robotique, Ecole des Mines de Paris (2003)
8. Devillers, O., Mourrain, B., Preparata, F., Trebuchet, P.: On circular cylinders by four or five points in space. Discrete and Computational Geometry 28, 83–104 (2003)
9. Durand, C.B.: Symbolic and numerical techniques for constraint solving. PhD thesis, Purdue University, Department of Computer Science, Major Professor-Christoph M. Hoffmann (1998)
10. Hoffmann, C.M., Yuan, B.: On spatial constraint solving approaches. In: Richter-Gebert, J., Wang, D. (eds.) ADG 2000. LNCS (LNAI), vol. 2061, pp. 1–15. Springer, Heidelberg (2001)
11. Kalkbrenner, M.: Solving systems of algebraic equations using Gröbner bases. In: Davenport, J.H. (ed.) ISSAC 1987 and EUROCAL 1987. LNCS, vol. 378, pp. 282–292. Springer, Heidelberg (1989)

12. Lazard, D., Rouillier, F.: Solving parametric polynomial systems. Journal of Symbolic Computation 42(6), 636–667 (2007)
13. Lichtblau, D.: Cylinders through five points: computational algebra and geometry (2006), http://www.math.kobe-u.ac.jp/icms2006/icms2006-video/video/v001.html
14. Lichtblau, D.: Simulating perturbation to enumerate parametrized systems. In: Manuscript (2006)
15. Macdonald, I.G., Pach, J., Theobald, T.: Common tangents to four unit balls in $\mathbb{R}^3$. Discrete and Computational Geometry 26(1), 1–17 (2001)
16. Manubens, M., Montes, A.: Improving the DISPGB algorithm using the discriminant ideal. Journal of Symbolic Computation 41, 1245–1263 (2006)
17. Manubens, M., Montes, A.: Minimal canonical comprehensive Gröbner systems. In: ArXiv Mathematics e-prints (2007)
18. Megyesi, G.: Lines tangent to four unit spheres with coplanar centres. Discrete and Computational Geometry 26(4), 493–497 (2001)
19. Mishra, B.: Algorithmic algebra. Springer, New York (1993)
20. Montes, A., Recio, T.: Automatic discovery of geometry theorems using minimal canonical comprehensive groebner systems. In: ArXiv Mathematics e-prints (2007)
21. Rusin, D.: The Mathematical Atlas, http://www.math-atlas.org/98/5pt.cyl
22. Schömer, E., Sellen, J., Teichmann, M., Yap, C.: Smallest enclosing cylinders. Algorithmica 27, 170–186 (2000)
23. Sturmfels, B.: Polynomial equations and convex polytopes. American Mathematical Monthly 105(10), 907–922 (1998)
24. Weispfenning, V.: Comprehensive Gröbner bases. Journal of Symbolic Computation 14(1), 1–29 (1992)

Detecting All Dependences in Systems of Geometric Constraints Using the Witness Method

Dominique Michelucci and Sebti Foufou

LE2I UMR CNRS 5158, UFR Sciences, Université de Bourgogne,
BP 47870, 21078 Dijon Cedex, France
dmichel@u-bourgogne.fr, sfoufou@u-bourgogne.fr

Abstract. In geometric constraints solving, the detection of dependences and the decomposition of the system into smaller subsystems are two important steps that characterize any solving process, but nowadays solvers, which are graph-based in most of the cases, fail to detect dependences due to geometric theorems and to decompose such systems. In this paper, we discuss why detecting all dependences between constraints is a hard problem and propose to use the witness method published recently to detect both structural and non structural dependences. We study various examples of constraints systems and show the promising results of the witness method in subtle dependences detection and systems decomposition.

1   Introduction

Today all CAD CAM geometric modelers provide a geometric solver that enables designers to define shapes (geometric configurations) as solutions of a set of geometric constraints [3,24,11,8]. Geometric constraints specify distances, angles, incidences, and tangencies between basic geometric elements such as points, lines, circles, conics or higher degree curves (*e.g.* Bézier curves) in 2D, and lines, planes, quadrics or higher degrees algebraic curves and surfaces in 3D. In practice, designers interactively specify constraints on an approximation of the wanted configuration (called a "sketch") – the solver is often called a sketcher. The solver operates in various steps: (i) reads the sketch; (ii) translates the system of constraints into some internal data structure (typically some graph, and a system of equations...); (iii) analyses and decomposes the system; (iv) solves the subsystems obtained from the decomposition either with some formula or with a numerical method; (v) and finally assembles solutions of subsystems and displays the corrected sketch.

As the system is typically non linear, there is usually more than one solution, and the solver is supposed to provide the solution that gives the closest configuration to the intention of the designer. It turns out that, in 90% of the cases, the Newton-Raphson method converges to this solution when it starts from the initial guess provided by the sketch. When the Newton-Raphson method fails, the designer can resort to another method, slower but safer, like homotopy: for

F. Botana and T. Recio (Eds.): ADG 2006, LNAI 4869, pp. 98–112, 2007.

an algebraic system of equations, the attraction basins for homotopy are semi algebraic sets, the ones of Newton-Raphson method are fractals; this is an empirical argument that the homotopy method should converge to the closest root more often than the Newton-Raphson method.

Dimensioning a complex mechanical part involves hundreds or thousands of geometric constraints, that is why qualitative analysis of the system of constraints plays an essential role in preparing the resolution [21,15,20,24]. Today this analysis seems to be graph-based for all industrial solvers, as far as it is possible to know.

Graph-based methods develop some kinds of graph representing the system of constraints; they compute the so called degrees of freedom in this graph and its subgraphs. Technically, graph-based methods compute maximum matching [23,1,10], or maximum flows [13,12], or k-connected components [17,20,21]. These methods are polynomial time; they work very well for correct systems of constraints, *i.e.* when constraints are independent. Indeed, graph-based methods allow to solve systems of constraints which could not be solved otherwise. Graph-based methods can also detect the simplest dependences between constraints, called structural dependences which typically occur when a subset of unknowns is constrained by too much constraints, as in the system $f(x, y, z) = g(z) = h(z) = 0$ which over constrains z.

It is essential to detect dependences because numeric solvers typically fail, or get bogged down, when they are used to solve systems which are "wrongly" assumed to be well-constrained (to have a finite number of roots modulo the group of isometries). Moreover they do not give any useful explanation to help the users fix the problem. However, when the system of constraints is available without any further details, no polynomial time method can detect all dependences. Non structural dependences, which are due to geometric theorems, are not detectable by the previous methods. These dependences can occur in the seemingly simplest geometric constraints such as point-line incidences in 2D (in the projective plane, more technically). In CAD CAM, the major part of systems of geometric constraints involve such incidence constraints, *i.e.* incidence relations between points, lines, planes, circles/spheres, conics/quadrics. Of course, other constraints are also used to specify angles and lengths for dimensioning; these metric constraints involve parameters (values for lengths and angles) with generic values. Incidence constraints are especially relevant; Section 3 shows that detecting non structural dependences just amongst point-line incidences in 2D is as difficult as the ideal or radical membership problem of computer algebra. This problem is decidable with standard bases (also called Grobner bases) computable with Buchberger's method [25,4]. But no method to solve it is practicable, *i.e.* none scales to problems with industrial size.

This difficulty explains why the GCS (Geometric Constraints Solving) community usually assumes that constraints are either independent, or structurally dependent. This inability to detect and treat non structural dependences clearly restricts the use of GCS. This paper shows that at least for CAD CAM problems, an alternative method is indeed able to detect all dependences, structural or non structural, in polynomial time, assuming that a witness (to be defined below in Section 2) is available. Moreover, this method can also be used to decompose a well-constrained system into smaller well constrained parts, also in polynomial time. Here, well-constrained means: has a finite number of solutions modulo some group; the most relevant group for dimensioning a part in CAD CAM is the group of isometries: all compositions of translations, rotations and symmetries. For short, a system or a subsystem which is "well constrained modulo the isometry group" is said to be rigid in the GCS community (the meaning of "rigid" is a bit different in the rigidity theory community which has inspired several graph-based methods used in GCS).

Section 2 defines the witness configuration, discusses the witness computation, and presents the probabilistic graph rigidity test of which the witness method is an offspring. Section 3 explains why detecting all dependences between constraints is as hard as the ideal or radical membership problem. Section 4 explains the rigidity test of the witness method. Section 5 gives the proof that the witness method detects all dependences, structural as well as non structural. Section 6 presents a possible method to decompose a rigid systems into rigid subsystems. Section 7 concludes.

2   The Witness Method

2.1   The Witness: Definition and Computation

A witness is defined as follows: let $F(U, X) = 0$ be the system to be solved; X is the vector of unknowns, and U is the vector of parameters (lengths, cosines, or sines, or tangents of angles, etc) or non geometric parameters (Young elasticity modules, weights, costs, densities, temperatures, forces, etc) [19]. By definition, the values of parameters are known just before the resolution. The goal is to find the roots X_T of the target system: $F(U_T, X_T) = 0$, where U_T are the specified values for the parameters U. Thus the target is a couple (U_T, X_T) so that $F(U_T, X_T) = 0$. A witness is just another couple (U_W, X_W) such that $F(U_W, X_W) = 0$ as well. Usually, U_W and U_T are different, so a witness root is likely not a target root; nevertheless, the witness and the target share essential combinatorial properties *e.g.* they share the same jacobian rank and structure. Actually, the witness method assumes that the target and the witness have the same combinatorial properties, in other words, only their numeric values are different. This is a probabilistic assumption in the following sense: among all possible witnesses, *i.e.* solutions of the system $F(U, X) = 0$, the set of wrong witnesses has measure (or probability) zero. A witness is wrong when its combinatorial properties (rank and structure of its jacobian) are different from those of the target. The principle of the witness method is then straightforward: it studies the combinatorial properties at the witness and transfers them to the target.

In CAD CAM, a witness is usually available, or easy to find; often the sketch is a possible witness. A witness is a configuration which fulfils all "constraints without parameters". These constraints are projective constraints of linear incidences (collinearities, coplanarities) or non linear incidences ("co-conicity" of 6 points in 2D), parallelisms or orthogonalities (or other non generic angles). In passing, parallelisms and orthogonalities (and non generic angles) can all be replaced by projective constraints, see Fig. 1, 2, and 3 for an intuitive account. The set of possible witnesses is very large: typically it is a continuum with dimension $|U|$ where $|U|$ is the number of parameters. This explains why, in the CAD CAM context it is generally easy to find a witness: the sketch, or the solution to a previously solved system is usually a good witness. However, the witness computation may also be arbitrarily difficult, We summarize here some cases where the difficulty is known. For the molecule problem (given some inter atomic distances, find the configuration of the molecule) [9,10,6], finding a witness is completely trivial: just generate random points in 2D or in 3D according to the nature of the problem. For Eulerian polyhedra with specified coplanarities constraints (metric constraints such as angles between planes, distances between points, are dismissed), finding a witness is cubic time; it results from a constructive proof of Steinitz's theorem [22], which states that each Eulerian polyhedron is realizable in $\mathbb{Z}^3$ with some convex polyhedron, *i.e.* all vertices coordinates are integers (Steinitz's property is remarkable, since it does not hold in 4D [22]). Fewer details are known for non Eulerian polyhedra, *i.e.* with one or several handles.

For general geometric constraints (including incidences), we can often use a dual method for generating a witness, for instance when the unknown configuration is a 3D polyhedron described by the length of its edges. The octahedron problem, solved by Durand and Hoffmann [7], also known as the Stewart platform, and the icosahedron problem are just molecule problems: they have triangular faces, so generating random vertices is sufficient; the fact that the generated polyhedron is likely concave and even self intersecting does not matter as far as the distance and coplanarity conditions are satisfied. The hexahedron (6 quadrangular planar faces) or the dodecahedron (12 pentagonal planar faces) are examples of systems of geometric constraints where the dual method for generating the witness works: generate random planes in 3D, one random plane per face, before computing the resulting vertices as intersection points of the supporting planes. This dual method clearly relies on the fact that each vertex of the hexahedron and of the dodecahedron is degree 3. For vertices with greater degree, the method will not work because there is a null probability for four (or more) random planes to meet in a common point.

2.2   A Forerunner of the Witness Method

The witness method extends a probabilistic test used in rigidity theory [9]. Rigidity theory searches a combinatorial characterization for the rigidity of graphs: for a given dimension, does a non oriented graph with edges labelled with generic lengths has a rigid realization in d dimensions? For instance two triangles sharing a common edge are a rigid graph in 2D, but not in 3D where they can fold along their common edge. The genericity assumption forbids collinearities, coplanarities in 3D, and non linear incidences (points on conics, or quadrics) which are essential in CAD CAM.

The characterization of graph rigidity is known in 2D, after Laman's theorem: a graph with $v \geq 2$ vertices and e edges is minimally rigid (or isostatic: it is rigid, and removing one edge makes it flexible) iff $e = 2v - 3$ and if for all subgraphs induced by v' vertices and having e' edges, $e' \leq 2v' - 3$. There is an exponential number of subgraphs, but several polynomial time algorithms have been proposed to test graph rigidity [18,10].

The intuitive extension to 3D of Laman's theorem (where $2v - 3$ is replaced by $3v - 6$, and $2v' - 3$ is replaced by $3v' - 6$) is unfortunately wrong; the double banana is the most famous example. Up to now, the combinatorial characterization of graph rigidity in 3D and beyond is unknown. For GCS, it is convenient to extend Laman's theorem to other kinds of constraints in 2D, and in 3D. It gives an *approximate* but essential characterization of rigidity on which graph-based decomposition methods rely.

Though the combinatorial characterization of graph rigidity is unknown in 3D and beyond, there is a probabilistic and polynomial time algorithm to decide the rigidity of a given graph for any given dimension. It relies on Gluck's theorem: a graph is rigid if a generic realization of it is. So compute the rank of the jacobian (called the rigidity matrix) for a random realization. The witness method is an offspring of this rigidity test; in order to account for any kind of constraints (and not only point-point distances), the realization can no more be random: a random realization has probability 0 to fulfil "constraints without parameters": parallelism, orthogonalities, incidences (collinearities, coplanarities).

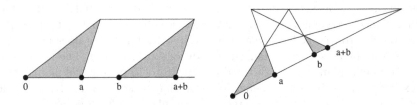

Fig. 1. Affine and projective construction of $a + b$. In the affine construction (left), the two shaded triangles are congruent and have parallel sides.

From a theoretical viewpoint, the witness method brings nothing new: it does not give a combinatorial characterization for well-constrainedness modulo the isometries group. It is a rather straightforward extension of the probabilistic test for graph rigidity which relies on Gluck theorem.

3   Why Detecting All Dependences Is Difficult

This section explains why a polynomial time method can not detect all dependences in seemingly simple systems of constraints containing only point-line incidences in 2D.

All systems of algebraic equations with coefficients in $\mathbb{Z}$ reduce, in polynomial time, to a system of point-line incidences in the projective plane [5,2]. The idea

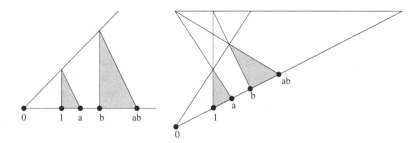

Fig. 2. Affine and projective construction of $a \times b$. In the affine construction, the two shaded triangles are similar (proportional) and have parallel sides.

is to represent each number (coefficient or root) by a point on an arbitrary line; this line passes through two arbitrary distinct points, representing the number 0 and the number 1 (later, the projective construction will need a third arbitrary point on this line, the point at infinity). Then a first geometric construction in 2D constructs the point representing $a + b$ from the point representing a and the point representing b , see Fig. 1. A second geometric construction constructs the point representing $a \times b$ from the points representing a and representing b, see Fig. 2. Actually, a construction in affine geometry is first proposed; it uses parallelism constraints, which are removed in the projective construction, using the classical idea of projective geometry: parallel lines are replaced by lines concurrent to a special arbitrary line, the Desargues line at infinity. It is a classical result that, if the projective plane satisfies Desargues and Pappus properties, then these two geometric constructions indeed define a field of numbers, *i.e.* associativity, commutativity, distributivity, etc hold; for instance the fact that point representing $a \times b$ is equal to the point $b \times a$ relies on Pappus property of the projective plane [5].

These two constructions permit to translate the algebraic system into a set of point-line incidences. Remark that the ruler construction of an integer coefficient n of the system of equations needs $O(\log_2 n)$ incidence constraints, using iterated

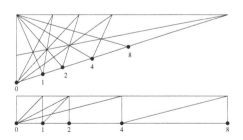

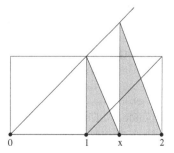

Fig. 3. Left: geometric construction of powers of 2. Right: affine constraints equivalent to the equation $x^2 - 2 = 0$.

squaring and additions (Fig. 3), thus it has the same size as the usual binary representation of integers. The right part of Fig. 3 shows the set of incidence and parallelism constraints for the equation $x^2 - 2 = 0$.

Thus solving algebraic systems and solving point-line incidences constraints are in principle the same problem. So detecting dependences in point-line incidences constraints is as difficult as detecting dependences in algebraic systems. The latter problem is equivalent to solving the ideal membership problem, and the radical membership problem (more on this question in section 3). Both problems are decidable, for instance with standard bases computable with Buchberger's algorithm. Unfortunately, due to the high algorithmic complexity of the problem, these methods are practicable only for small instances. Another direct consequence is that, in principle, finding a witness can be arbitrarily difficult: it suffices to translate a difficult system of equations (*e.g.* arising from a system of geometric constraints with specified numerical values for parameters) into a system of point-line incidences in 2D.

4 The Study of a Witness

This section details the study of the witness. The main idea is to compute the structure of the jacobian of the system at the witness. For example, if the jacobian has full rank, then the witness system contains no dependence, and this property is transferred to the target system.

The kernel of the jacobian at the witness is a vector space of the so called infinitesimal motions. Classically, there are two kinds of motions: displacements (also called rigid motions, or isometries), and flexions. Displacements are compositions of translations, symmetries, rotations; they constitute the isometry group; they do not alter distances and angles. On the contrary, flexions do – at least in the generic case (degenerate cases are not considered in this paper, for concision).

4.1 Computing a Basis of Infinitesimal Displacements

A system of geometric constraints is rigid iff the kernel of its jacobian contains only displacements. It is possible to compute an *a priori* basis of the infinitesimal displacements. Table 1 shows such a basis, in the 2D case, composed of t_x a translation in the x direction, t_y a translation in the y direction, and r_{xy} a rotation around the origin. (x_i, y_i) are coordinates of a point, (a_l, b_l, c_l) are coordinates of a line (*i.e.* the line has equation: $a_l x + b_l y + c_l = 0$), and (u_k, v_k) are coordinates of a vector (the difference between 2 points). q_j represents an unknown which is independent of the cartesian frame: it is either a geometric unknown such as a length, a radius, an area, a scalar product, or a non geometric unknown. Dotted variables $\dot{x}_i, \dot{y}_i, \dot{a}_l, \dot{b}_l, \dot{c}_l, \dot{u}_k, \dot{v}_k$ and $\dot{q}_j$ are used to denote the values of the corresponding coordinates in the basis of infinitesimal displacements, *e.g.* the couple

Table 1. A basis for the free displacements in 2D for points, lines, and vectors

	$\dot{x}_i$	$\dot{y}_i$	$\dot{a}_l$	$\dot{b}_l$	$\dot{c}_l$	$\dot{u}_k$	$\dot{v}_k$	$\dot{q}_j$
t_x	1	0	0	0	$-a_l$	0	0	0
t_y	0	1	0	0	$-b_l$	0	0	0
r_{xy}	$-y_i$	x_i	$-b_l$	a_l	0	$-v_k$	u_k	0

$(\dot{x}_i, \dot{y}_i)$ representing the infinitesimal translation t_x along the x axis of a point (x_i, y_i) is equal to $(1, 0)$. Note that the infinitesimal displacements for a point (x, y), a normal (a, b) to a line, and a vector (u, v) are different; *e.g.* translating a point modify it, but translating a vector or a normal does not.

Table 2 shows a possible basis for the infinitesimal displacements in 3D; it contains three translations t_x, t_y, and t_z, and three rotations r_{xy}, r_{xz} and r_{yz}. Points have coordinates (x, y, z), planes have coordinates (a, b, c, d) (their equation is: $ax + by + cz + d = 0$), vectors have coordinates (u, v, w); q represents an unknown independent of the cartesian frame.

Table 2. A basis for the free displacements in 3D for points, planes, vectors, and unknowns independent of the cartesian frame

	$\dot{x}_i$	$\dot{y}_i$	$\dot{z}_i$	$\dot{a}_h$	$\dot{b}_h$	$\dot{c}_h$	$\dot{d}_h$	$\dot{u}_k$	$\dot{v}_k$	$\dot{w}_k$	$\dot{q}_j$
t_x	1	0	0	0	0	0	$-a_h$	1	0	0	0
t_y	0	1	0	0	0	0	$-b_h$	0	1	0	0
t_z	0	0	1	0	0	0	$-c_h$	0	0	1	0
r_{xy}	$-y_i$	x_i	0	$-b_h$	a_h	0	0	$-v_k$	u_k	0	0
r_{xz}	$-z_i$	0	x_i	$-c_h$	0	a_h	0	$-w_k$	0	u_k	0
r_{yz}	0	$-z_i$	y_i	0	$-c_h$	b_h	0	0	$-w_k$	v_k	0

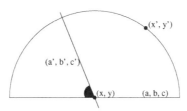

Fig. 4. A 2D under-constrained system of geometric constraints

4.2   A Structurally Under-Constrained Example in 2D

Fig. 4 shows a simple under-constrained example in 2D. A possible witness is $(x = y = 0, x' = 3, y' = 4, \delta = 5, a = 1, b = 0, a' = 12/13, b' = 5/13, \text{ and } \lambda = 12/13)$. All graph-based methods give correct results when considering this system,

Table 3. The jacobian and a basis of infinitesimal motions: three displacements and a flexion for the system given in (1)

	x	y	x'	y'	a	b	c	a'	b'	c'
e_1'	a	b	0	0	x	y	1	0	0	0
e_2'	a'	b'	0	0	0	0	0	x	y	1
e_3'	$2(x-x')$	$2(y-y')$	$2(x'-x)$	$2(y'-y)$	0	0	0	0	0	0
e_4'	0	0	0	0	$2a$	$2b$	0	0	0	0
e_5'	0	0	0	0	0	0	0	$2a'$	$2b'$	0
e_6'	0	0	0	0	a'	b'	0	a	b	0
	$\dot{x}$	$\dot{y}$	$\dot{x}'$	$\dot{y}'$	$\dot{a}$	$\dot{b}$	$\dot{c}$	$\dot{a}'$	$\dot{b}'$	$\dot{c}'$
t_x	1	0	1	0	0	0	$-a$	0	0	$-a'$
t_y	0	1	0	1	0	0	$-b$	0	0	$-b'$
r_{xy}	$-y$	x	$-y'$	x'	$-b$	a	0	$-b'$	a'	0
flexion	0	0	$y-y'$	$x'-x$	0	0	0	0	0	0

because the under-constrainedness is structural. This system is composed of the following six equations:

$$e_1 : ax + by + c = 0$$
$$e_2 : a'x + b'y + c' = 0$$
$$e_3 : (x - x')^2 + (y - y')^2 - \delta^2 = 0 \qquad (1)$$
$$e_4 : a^2 + b^2 - 1 = 0$$
$$e_5 : a'^2 + b'^2 - 1 = 0$$
$$e_6 : aa' + bb' - \lambda = 0$$

The jacobian and a basis of its kernel are given Table 3 where all symbols are replaced by their values at the witness. This basis contains the 3 displacements of the plane, plus a flexion vector: indeed point (x', y') can rotate around point (x, y). If columns x', y' are removed, the flexion vector clearly becomes, in the remaining columns: $x, y, a, b, c, a', b', c'$, a linear combination of the 3 displacement vectors (the fact that it vanishes is basis dependent; the fact that it is a linear combination is not). This shows that the part obtained after the removal of the point (x', y') is rigid. This is the test that the witness method uses to decide the rigidity of a part.

4.3   A Dependence due to a Theorem

Let us now consider the 2D system of geometric constraints of Fig. 5. This system is structurally correct, but contains a non structural dependence due to a (simple) geometric theorem, so it will be difficult (at least) for graph-based methods to detect this dependence, but the witness method detects it.

Constraints are as follows; point O is the middle of AB; distance OC equals distance OA; the angle between AC and BC is right. Actually this constraint is a consequence of the other constraints. Finally distance OA is specified (just to have the "good" number of constraints). The system of equations is:

$$e_1 : 2x_O - x_A - x_B = 0$$
$$e_2 : 2y_O - y_A - y_B = 0$$
$$e_3 : (x_C - x_O)^2 + (y_C - y_O)^2 - (x_A - x_O)^2 - (y_A - y_O)^2 = 0 \qquad (2)$$
$$e_4 : (x_C - x_A)(x_C - x_B) + (y_C - y_A)(y_C - y_B) = 0$$
$$e_5 : (x_A - x_O)^2 + (y_A - y_O)^2 - u^2 = 0$$

A possible witness is: $O = (0,0), A = (-10,0), B = (10,0), C = (6,8), u = 10$. The jacobian and a basis of the free infinitesimal motions (three displacements and a flexion: point C can rotate around point O) are given in Table 4, where again, all symbols are replaced by their value at the witness. The rank of $e'_1, \ldots, e'_5$ computed at the witness is 4, thus equations are dependent.

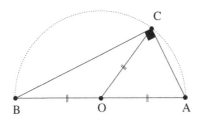

Fig. 5. Example of dependent constraints

Table 4. The jacobian, and a basis of 4 free infinitesimal motions for the dependent system given in (2). The fourth motion is a flexion: point C can rotate around O.

	x_O	y_O	x_A	y_A	x_B	y_B	x_C	y_C
e'_1	2	0	-1	0	-1	0	0	0
e'_2	0	2	0	-1	0	-1	0	0
e'_3	$2x_A - 2x_C$	$2y_A - 2y_C$	$2x_O - 2x_A$	$2y_O - 2y_A$	0	0	$2x_C - 2x_O$	$2y_C - 2y_O$
e'_4	0	0	$x_B - x_A$	$y_B - y_C$	$x_A - x_C$	$y_A - y_C$	$2x_C - x_A - x_B$	$2y_C - y_A - y_B$
e'_5	$2x_O - 2x_A$	$2y_O - 2y_A$	$2x_A - 2x_O$	$2y_A - 2y_O$	0	0	0	0

	x_O	y_O	x_A	y_A	x_B	y_B	x_C	y_C
t_x	1	0	1	0	1	0	1	0
t_y	0	1	0	1	0	1	0	1
r_{xy}	$-y_O$	x_O	$-y_A$	x_A	$-y_B$	x_B	$-y_C$	x_C
flexion	0	0	0	0	0	0	$y_O - y_C$	$x_C - x_O$

4.4   Computing Degrees of Displacements

In an attempt to make graph-based methods more robust against non structural dependences, Jermann introduced the notion of DoD, Degrees of Displacements, in his PhD thesis (actually he called that the Degree of Rigidity) [16,14]. Consider a subset of unknowns $Y \subset X$; extract the columns of Y in the basis of the infinitesimal displacements; call it $D[Y]$; then the DoD of Y is the rank of this subarray $D[Y]$. For instance, for a 2D line $Y = (a, b, c)$, the subarray $D[Y]$ is: $(\dot{a}, \dot{b}, \dot{c}) = (0, 0, -a)$ for the translation t_x, $(0, 0, -b)$ for the translation t_y, and $(-b, a, 0)$ for the rotation r_{xy}. The line has 2 DoD and the 2 translations are dependent. Similarly, $M[Y]$ will denote the content of the columns Y in any basis of the free infinitesimal motions (displacements, and flexions) of the system at the witness.

Jermann understood that graph-based methods are fragile because they can not compute exact DoDs. It turns out that DoD is not only a syntactical or structural idea when accounting for incidence constraints. For instance, the DoD of two secant planes in 3D is 5 (more precisely the 3 rotations are independent, but the 3 translations are dependent; intuitively, and *a posteriori*, it is understandable, since translating the 2 planes along their intersection line leave them invariant) while the DoD of two parallel planes is 4. Similarly, the DoD of 3 collinear points is 2, while the DoD of 3 non collinear points is 3. A pure graph-based method has no mean to know if 3 points are collinear or not, or if two planes are parallel or not. Either it assumes the configuration is generic (and thus has maximal DoD), or it can try to look if the parallelism/collinearity is an explicit constraint of the system; but it may happen that the parallelism/collinearity is a remote consequence of a set of constraints, thanks to Desargues, or Pappus, or Pascal, or Miquel theorems: the incidence in the conclusion is a non trivial consequence of the hypothesis.

The witness method avoids this difficulty: it just computes the DoD of the part, at the witness, with standard method from linear algebra. If a parallelism/collinearity or any other feature holds because of some constraints and geometric theorems, then it holds in the witness.

A part with full DoD (3 in 2D, 6 in 3D) and with minimal cardinality: 3 in 2D, 6 in 3D, is called an anchor. Section 6 uses anchors for decomposing.

4.5   Interrogating a Witness

Geometric constraints are independent of the cartesian frame. But sometimes, some constraints such as: $x_1 = y_1 = y_2 = 0$ are used to "pin" the configuration in the plane and to make the system of equations well-constrained, which simplifies the work of the numerical solver, *e.g.* Newton-Raphson. The following test checks that a constraint is independent of the cartesian frame: it is iff its gradient vector is orthogonal to all basis vectors in the basis of infinitesimal displacements. From now on, constraints are assumed to be independent of the cartesian frame.

Are constraints independent? They are iff the jacobian at the witness has full rank. Is a part Y rigid, *i.e.* well constrained modulo displacements? It is iff $D[Y] \equiv M[Y]$: the vector space of its infinitesimal motions is equal to the vector space of its infinitesimal displacements. In other words, since $D[Y] \subset M[Y]$, the rank of $D[Y]$ is equal to the rank of $M[Y]$.

4.6   Rank Computations Are All What You Need

All computations of ranks are performed on vectors with numerical entries provided by the coordinates of the witness and the gradient vectors at the witness.

Inaccuracy issue is detailed elsewhere. We just mention that, if the witness has rational coordinates, then all computations can be performed exactly, with the classical Gauss pivoting method; it is also possible to compute exactly modulo a prime close to 10^9, which introduces another source of probabilisticity. Finally, if the witness method is used only to check the independence of vectors, then an interval arithmetic is sufficient, in the following sense: if a set of vectors is independent, then interval computations can guarantee it (assuming intervals are sharp enough); if vectors are dependent, then interval computations can not prove it. Note that when the interval analysis can not prove the independence of the gradient vectors at the witness, either vectors are dependent or the system at the witness is ill-conditioned.

5   The Proof That the Witness Method Detects All Dependences

An algebraic equation $g(x) = 0$ is a consequence of the other equations: $f_1(x) = f_2(x) = \ldots f_n(x_n) = 0$ in two cases: either g lies in the ideal generated by the polynomials $f_1, \ldots f_n$, or g lies in the radical generated by the polynomials $f_1, \ldots f_n$. This section proves that in both cases the witness method detects the dependence.

Assume first that g lies in the ideal of the polynomials $f_i, i = 1, \ldots n$. Then by definition there are polynomials $\lambda_1(x), \ldots \lambda_n(x)$ such that $g(x) = \lambda_1(x) \times f_1(x) + \lambda_2(x) \times f_2(x) + \ldots \lambda_n(x) \times f_n(x)$. A first consequence is that g vanishes at a common root of the polynomials $f_i, i = 1, \ldots n$. A second consequence is obtained by deriving the previous equality: $\nabla g(x) = \nabla \lambda_1(x) f_1(x) + \lambda_1(x) \nabla f_1(x) + \ldots \nabla \lambda_n(x) f_n(x) + \lambda_n(x) \nabla f_n(x)$. At a common root w of the f_i polynomials, such as the witness, terms $f_i(w)$ vanish and it yields $\nabla g(w) = \lambda_1(w) \nabla f_1(w) + \ldots \lambda_n(w) \nabla f_n(w)$. In other words the gradient vector of $g(w)$ is a linear combination of the gradient vectors $f_1(w), \ldots f_n(w)$. But the witness method detects such dependences between the gradient vectors when it studies the jacobian at the witness w.

Assume now that g does not lie in the ideal, but in the radical of $f_1, \ldots f_n$. By definition there is an integer $k > 1$ and polynomials $\lambda_1(x), \ldots \lambda_n(x)$ such that $g(x)^k = \lambda_1(x) \times f_1(x) + \lambda_2(x) \times f_2(x) + \ldots \lambda_n(x) \times f_n(x)$. A first consequence is that g vanishes at a common root of the polynomials $f_i, i = 1, \ldots n$. A second consequence is that, by derivation: $kg(x)^{k-1} \nabla g(x) = \nabla \lambda_1(x) f_1(x) + \lambda_1(x) \nabla f_1(x) + \ldots \nabla \lambda_n(x) f_n(x) + \lambda_n(x) \nabla f_n(x)$. At a common root w of the f_i polynomials, terms $f_i(w)$ vanish and it yields $kg(w)^{k-1} \nabla g(w) = 0 = \lambda_1(w) \nabla f_1(w) + \ldots \lambda_n(w) \nabla f_n(w)$. It means that the gradient vectors of $f_1, \ldots f_n$ at w are dependent (and g does not matter in this case). But the witness method detects such dependences between the gradient vectors when it studies the jacobian at the witness w. Thus, in both cases, the witness method detects a dependence between the gradient vectors of the equations at the witness. The previous proof applies to algebraic systems. Maybe it is possible to extend it to non algebraic equations (involving transcendentals) with some topological argument.

6   The Witness Method Decomposes into Rigid Parts

Every graph-based method for decomposing the system of geometric constraints and assembling the solutions of the parts, is two-folds: on one hand it proposes a strategy to decompose the system, and on the other hand it proposes a test to decide if a given part is well-, over-, or under-constrained modulo the isometries group. We saw that, contrary to the witness method, the test can be confused with some configurations and fail to detect some dependences when they are not structural. But the strategy can still be used. Thus for each graph-based method proposed so far to plan the resolution process, it is possible to keep its strategy, and to replace its rigidity test with the one provided by the witness method.

This section explains one of the possible strategies to decompose a rigid system into rigid subsystems, maybe the simplest strategy. If the system is flexible, its Maximal Rigid Parts (MRP) are computed. If it is rigid, each constraint is removed in turn; it provides a flexible system, the MRP of which are computed.

To find the MRP of a flexible system, its anchors are first determined; an anchor is a subset Y of $d(d+1)/2$ unknowns which has full DoD (3 in 2D, 6 in 3D) and which is rigid, $i.e.$ the vector space of its infinitesimal motions $M[Y]$ is equal to the vector space of its infinitesimal displacements $D[Y]$. Clearly there is a polynomial number of potential anchors; just test the rigidity of each potential anchor. Every anchor Y belongs to exactly one MRP, noted MRP(Y). MRP(Y) is computed with the obvious greedy method: initialize MRP(Y) with Y, consider every variable $x \in X - Y$ (in any order) and insert it in MRP(Y) iff $Y \cup \{x\}$ is still rigid, $i.e.$ iff $M[Y \cup \{x\}] \equiv D[Y \cup \{x\}]$.

Some book-keeping may speed up the method, and avoid to find several times the same maximal rigid parts. However the method is polynomial even without such optimizations: indeed there is a polynomial number of potential anchors, and each anchor is contained in a single MRP.

7   Conclusion

This paper has shown that the witness method detects all dependences: structural dependences which are already detected by graph-based methods, but also non structural dependences which are due to known or unknown geometric theorems, and may occur with incidence constraints. The witness method can also decompose a rigid system into rigid subsystems; actually it is possible to reuse the strategic part of every graph-based method proposed so far to decompose rigid systems into rigid irreducible parts with the rigidity test provided by the witness method. In practice, the witness method should widen the scope of geometric constraints solving.

References

1. Ait-Aoudia, S., Jegou, R., Michelucci, D.: Reduction of constraint systems. In: Compugraphic, Alvor, Portugal, pp. 83–92 (1993)
2. Bonin, J.E.: Introduction to matroid theory. The George Washington University web site (2001)
3. Bruderlin, B., Roller, D. (eds.): Geometric Constraint Solving and Applications. Springer, Heidelberg (1998)
4. Cox, D., Little, J., O' Shea, D.: Ideals, varieties and algorithms, 2nd edn. Springer, Heidelberg (1996)
5. Coxeter, H.: Projective geometry. Springer, Heidelberg (1987)
6. Crippen, G.M., Havel, T.F.: Distance Geometry and Molecular Conformation, Taunton, U.K. Research Studies Press (1988) ISBN 0-86380-073-4
7. Durand, C., Hoffmann, C.M.: Variational constraints in 3D. In: Shape Modeling International, pp. 90–97 (1999)
8. Gao, X.-S., Zhang, G.: Geometric constraint solving via c-tree decomposition. In: ACM Solid Modelling, pp. 45–55. ACM Press, New York (2003)
9. Graver, J.E., Servatius, B., Servatius, H.: Combinatorial Rigidity. Graduate Studies in Math., AMS (1993)
10. Hendrickson, B.: Conditions for unique graph realizations. SIAM J. Comput. 21(1), 65–84 (1992)
11. Hoffmann, C.M.: Summary of basic 2D constraint solving. International Journal of Product Lifecycle Management 1(2), 143–149 (2006)
12. Hoffmann, C.M., Lomonosov, A., Sitharam, M.: Decomposition plans for geometric constraint problems, part II: New algorithms. J. Symbolic Computation 31, 409–427 (2001)
13. Hoffmann, C.M., Lomonosov, A., Sitharam, M.: Decomposition plans for geometric constraint systems, part I: Performance measures for CAD. J. Symbolic Computation 31, 367–408 (2001)
14. Jermann, C., Neveu, B., Trombettoni, G.: A new structural rigidity for geometric constraint systems. In: Winkler, F. (ed.) ADG 2002. LNCS (LNAI), vol. 2930, pp. 87–106. Springer, Heidelberg (2004)
15. Jermann, C., Trombettoni, G., Neveu, B., Mathis, P.: Decomposition of geometric constraint systems: a survey. Internation Journal of Computational Geometry and Applications (IJCGA) (to appear 2006)
16. Jermann, C., Trombettoni, G., Neveu, B., Rueher, M.: A new heuristic to identify rigid clusters. In: Richter-Gebert, J., Wang, D. (eds.) ADG 2000. LNCS (LNAI), vol. 2061, pp. 2–6. Springer, Heidelberg (2001)
17. Latham, R.S., Middletich, A.E.: Connectivity analysis: a tool for processing geometric constraints. Computer-Aided Design 28(11), 917–928 (1996)
18. Lovasz, L., Yemini, Y.: On generic rigidity in the plane. SIAM J. Algebraic Discrete Meth. 3(1), 91–98 (1982)
19. Michelucci, D., Foufou, S.: Geometric constraint solving: the witness configuration method. Computer Aided Design 38(4), 284–299 (2006)
20. Owen, J.C.: Algebraic solution for geometry from dimensional constraints. In: Proc. of the Symp. on Solid Modeling Foundations and CAD/CAM Applications, pp. 397–407 (1991)
21. Owen, J.C.: Constraint on simple geometry in two and three dimensions. Int. J. Comput. Geometry Appl. 6(4), 421–434 (1996)

22. Richter-Gebert, J.: Realization Spaces of Polytopes. LNCM, vol. 1643. Springer, Heidelberg (1996)
23. Serrano, D.: Automatic dimensioning in design for manufacturing. In: Sympos. on Solid Modeling Foundations and CAD/CAM Applications, pp. 379–386 (1991)
24. Sitharam, M.: Combinatorial approaches to geometric constraint solving: Problems, progress and directions. In: Dutta, D., Janardan, R., Smid, M. (eds.) AMS/DIMACS. Computer aided design and manufacturing (2005)
25. Wester, M.J.: Contents of Computer Algebra Systems: A Practical Guide. John Wiley and Sons, Chichester (1999)

Automatic Discovery of Geometry Theorems Using Minimal Canonical Comprehensive Gröbner Systems

Antonio Montes[1,*] and Tomás Recio[2,**]

[1] Dep. Matemàtica Aplicada 2, Universitat Politècnica de Catalunya, Spain
antonio.montes@upc.edu
http://www-ma2.upc.edu/~montes
[2] Dep. Matemáticas, Estadística y Computación, Universidad de Cantabria, Spain
tomas.recio@unican.es
http://www.recio.tk

Abstract. The main proposal in this paper is the merging of two techniques that have been recently developed. On the one hand, we consider a new approach for computing some specializable Gröbner basis, the so called Minimal Canonical Comprehensive Gröbner Systems (MCCGS) that is -roughly speaking- a computational procedure yielding "good" bases for ideals of polynomials over a field, depending on several parameters, that specialize "well", for instance, regarding the number of solutions for the given ideal, for different values of the parameters. The second ingredient is related to automatic theorem discovery in elementary geometry. Automatic discovery aims to obtain complementary (equality and inequality type) hypotheses for a (generally false) geometric statement to become true. The paper shows how to use MCCGS for automatic discovering of theorems and gives relevant examples.

Keywords: automatic discovering, comprehensive Gröbner system, automatic theorem proving, canonical Gröbner system.

MSC: 13P10, 68T15.

1 Introduction

1.1 Overview of Goals

The main idea in this paper is that of merging two recent techniques. On the one hand, we will consider a method (named *MCCGS*, standing for *minimal canonical comprehensive Gröbner systems*) [MaMo06], that is –roughly speaking– a

* Work partially supported by the Spanish Ministerio de Ciencia y Tecnología under project MTM 2006-01267, and by the Generalitat de Catalunya under project 2005 SGR 00692.
** Work partially supported by the Spanish Ministerio de Educación y Ciencia under project GARACS, MTM2005-08690-C02-02.

F. Botana and T. Recio (Eds.): ADG 2006, LNAI 4869, pp. 113–138, 2007.

computational approach yielding "good" bases for ideals of polynomials over a field depending on several parameters, where "good" means that the obtained bases should specialize (and specialize "well", for instance, regarding the number of solutions for the given ideal) for different values of the parameters.

Briefly, in order to understand what kind of problem $MCCGS$ addresses, let us consider the ideal $(ax, x + y)K[a][x, y]$, where a is taken as a parameter and K is a field. Then it is clear that there will be, for different values of $a = a_0 \in K$, essentially two different types of bases for the specialized ideal $(a_0 x, x+y)K[x, y]$. In fact, for $a_0 = 0$ we will get $(x + y)$ as a Gröbner-basis (in short, a G-basis) for the specialized ideal; and for any other rational value of a such that $a = a_0 \neq 0$, we will get a G-basis with two elements, (x, y). Thus, the given G-basis $(ax, x + y)K[a, x, y]$ does not specialize well to a G-basis of every specialized ideal. On the other hand, let us consider $(ax - b)K[a, b][x]$, where a, b are taken as free parameters and x is the only variable. Then, no matter which rational values a_0, b_0 are assigned to a, b, it happens that $\{a_0 x - b_0\}$ remains a Gröbner basis for the ideal $(a_0 x - b_0)K[x]$. Still, there is a need for a case-distinction if we focus on the cardinal of the solutions for the specialized ideal. Namely, for $a_0 \neq 0$ there is a unique solution $x = -b_0/a_0$; for $a_0 = 0$ and $b_0 \neq 0$ there is no solution at all; and for $a_0 = b_0 = 0$ a solution can be any value of x (no restriction, one degree of freedom).

The goal of $MCCGS$ is to describe, in a compact and canonical form, the discussion, depending on the different values of the parameters specializing a given parametric system, of the different basis for the resulting specialized systems and on their solutions.

The second ingredient of our contribution is about automatic theorem discovery in elementary geometry. Automatic discovery aims to obtain complementary hypotheses for a (generally false) geometric statement to become true. For instance, we can consider an arbitrary triangle and the feet on each sides of the three altitudes. These three feet give us another triangle, and now we want to conclude that such triangle is equilateral. This is generally false, but, under what extra hypotheses (of equality type) on the given triangle will it become generally true?

Finding, in an automatic way, the necessary and sufficient conditions for this statement to become a theorem, is the task of automatic discovery. A protocol for automatic discovery is presented in [RV99] and a detailed discussion of the method appears in [DR]. The protocol proceeds requiring some computations (contraction, saturation, etc.) about certain ideals built up from the given statement, but does not state any preference about how to perform such computations (although the computed examples in both papers rely on straightforward Gröbner bases computations for ideal elimination).

Our goal in this paper is to show how we can improve the automatic discovery of geometry theorems, by performing a $MCCGS$ procedure on an ideal built up from the given hypotheses and theses, considering as parameters the free coordinates of some elements of the geometric setting,

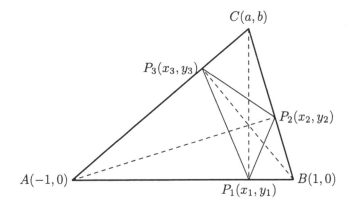

Fig. 1. Orthic triangle

1.2 Related Work

This idea has a close precedent in the work[1] of [CLLW], directly inspired by [K95] and, to a lesser extent, by [Weis92]. In [CLLW], a parametric radical membership test is presented for a mathematical construct the authors introduce, called "partitioned parametric Gröbner basis" (PPGB). Suppose we are given a statement $H := \{h_1 = 0, \ldots, h_r = 0\} \Rightarrow T := \{g = 0\}$, expressed in terms of polynomial equations –usually over some computable field– and their solutions over some extension field K –that we can assume, in order to simplify the exposition, to be algebraically closed. Roughly speaking, the method behind [CLLW] starts by computing the "partitioned basis" (with respect to a given subset of variables, here denoted by $\mathbf{u}$) of an ideal $I \subseteq K[\mathbf{u}, \mathbf{x}, y]$, (for instance, $I = (h_1(\mathbf{u}, \mathbf{x}) \ldots h_r(\mathbf{u}, \mathbf{x})$, $g(\mathbf{u}, \mathbf{x}) y - 1))$, ie. a finite collection of couples (C_i, F_i), where the C_i's are constructible sets described as $\{c_1 = 0, \ldots, c_m = 0, q_1 \neq 0, \ldots, q_s \neq 0\}$ on the parameter space, and the F_i's are some collections of polynomials in $K[\mathbf{u}, \mathbf{x}, y]$. Moreover, it is required (among other conditions) that the C_i's conform a partition of the parameter space and, also, that for every element $\mathbf{u_0}$ in each C_i, the (reduced) G-basis of $(h_1(\mathbf{u_0}, \mathbf{x}) \ldots h_r, (\mathbf{u_0}, \mathbf{x}), g(\mathbf{u_0}, \mathbf{x}) y - 1)$ is precisely $F_i(\mathbf{u_0}, \mathbf{x})$. It is well known (e.g. [K86] or [Ch88]) that, in this context, a statement $\{h_1 = 0 \ldots h_r = 0\} \Rightarrow \{g = 0\}$ is to be considered true if $1 \in (h_1 \ldots h_r, g\, y - 1)$; thus, the extra hypotheses that [CLLW] proposes to add for the statement to become a theorem are precisely those expressed by any of the C_i's such that the corresponding $F_i = \{1\}$, since this is the only case F_i can specialize to $\{1\}$.

We must remark that, simply testing for $1 \in (h_1 \ldots h_r, g\, y - 1)$, as in the method above, can yield to theorems that hold just because the hypotheses are not consistent (i.e. such that already $1 \in (h_1 \ldots h_r)$). This cannot happen with our approach to automatic discovery: if a new statement is discovered, then the obtained hypotheses will be necessarily consistent.

---

[1] But notice the authors of [CLLW] already mention the paper of Montes [Mo02] as a predecessor on this particular kind of discussion of Gröbner basis with parameters.

Although our approach stems from the same basic ideas, our contribution differs from [CLLW] in some respects: first, we focus on automatic discovery, and not in automatic proving. Moreover, we are able to specifically describe the capability and limitations of the method (while in [CLLW] it is only mentioned that, in the reducible case, their "method ... cannot determine if the conclusion of the geometric statement is true on some components of the hypotheses"). Second, even for proving, the use of *MCCGS* provides not only the specialization property (which is the key for the application of partitioned parametric bases in [CLLW]) but also an automatic case distinction, that allows a richer understanding of the underlying geometry for the considered situation. In fact, it seems that the partitioned parametric G-Basis (PPGB) algorithm from [CLLW] is close to the algorithm DISPGB considered in [Mo02], both sharing that their output requires collecting by hand multiple cases (and then having to manually express in some simplified way the union of the corresponding conditions on the parameters). Actually, the motivation for *MCCGS* was, precisely, improving DISPGB.

Our approach has also an evident connection (since [Weis92] is the common origin of all posterior developments on parametric Gröbner basis) to the work of several members of Prof. Weispfenning's group, regarding generic quantifier elimination (Q. E.) and its application to automatic theorem proving (as, for example, in [DG], [DSW], [SS], [St]). In particular we remark the strong relation of our work with that of [DG], that approaches theorem proving via a restricted (generically valid) Q.E. method, relying on generic Gröbner systems computations. The set of restrictions Θ provided by this method, besides speeding up the Q.E. computations, can be interpreted in the context of theorem proving, roughly speaking, as a collection of new (sufficient) non-degeneracy conditions for an statement to hold true.

Again, the difference between our contribution here and theirs is, first, that we address problems requiring, in general, parameter restrictions that go beyond "a conjunction Θ of negated equations in the parameters" ([DG], first paragraph in Section 3). That is, we deal with formulas that are almost always false (see below for a more detailed explanation of the difference between automatic derivation and automatic discovery) and require non-negated (ie. equality) parameter restrictions; they can not be directly approached via generic Q.E. since our formulas are, quite often, generically false. Moreover, our approach is limited to this specific kind of generically false problems and we do not intend to provide a general method for Q.E. A second difference is that, for our very particular kind of problems, *MCCGS* formulates parameter restrictions in a compact and canonical way, a goal that is not specifically intended concerning the description of Θ in [DG]. For these reasons we can not include performance comparisons to these Q.E. methods and we do not consider relevant (although we provide some basic information) giving hardware details, computing times, etc. on the performance of our method running on the examples described in the last section of this paper. We are not proposing something better, but something different in a different context.

Next Section includes a short introduction to the basics on automatic discovery, which could be of interest even for automatic proving practitioners. Section 3 provides some bibliographic references for the problem of the G-basis specialization and summarizes the main features of the *MCCGS* algorithm, including an example of its output. Section 4 describes the application of *MCCGS* to automatic discovery, while Section 5 works in detail a collection of curious examples, including the solution of a pastime from *Le Monde* and the simpler solution (via this new method) of one example also solved by a more traditional method.

2   A Digest on Automatic Discovery

Although less popular than automatic proving, automatic discovery of elementary geometry theorems is not new. It can be traced back to the work of Chou (see [Ch84], [Ch87] and [ChG90]), regarding the "automatic derivation of formulas", a particular variant of automatic discovery where the goal consists in deriving results that always occur under some given hypotheses but that can be formulated in terms of some specific set of variables (such as expressing the area of a triangle in terms of the lengths of its sides). Finding the geometric locus of a point defined through some geometric constraints (say, finding the locus of a point when its projection on the three sides of a given triangle form a triangle of given constant area [Ch88], Example 5.8) can be considered as another variant of this "automatic derivation" approach.

Although "automatic derivation" (or locus finding) aims to discover some new geometric statements (without modifying the given hypotheses), it is not exactly the same as "automatic discovery" (in the sense we have presented it in the previous section), that searches for complementary hypotheses for a (generally false) geometric statement to become true (such as stating that the three feet of the altitudes for a given triangle form an equilateral triangle and finding what kind of triangles verify it). Again, automatic discovery in this precise sense appears in the early work of Chou (whose thesis [Ch85] deals with "Proving and discovering theorems in elementary geometries using Wu's method") and Kapur [K89] (where it is explicitly stated that "...the objective here is to find the missing hypotheses so that a given conclusion follows from a given incomplete set of hypotheses...").

Further specific contributions to automatic discovery appear in [Wa98], [R98] (a book written in Spanish for secondary education teachers, with circa one hundred pages devoted to this topic and with many worked out examples), [RV99], [Ko] or [CW]. Examples of automatic derivation, locus finding and discovery, achieved through a specific software named *GDI* (the initials of *Geometría Dinámica Inteligente*), of Botana-Valcarce, appear in [BR05] or [RB] (and the references thereof), such as the automatic derivation of the thesis for the celebrated Maclane 8_3-Theorem, or the automatic answer to some items on a test posed by Richard [Ri], on proof strategies in mathematics courses, for students 14-16 years old.

The simple idea behind the different approaches is[2], essentially, that of adding the conjectural theses to the collection of hypotheses, and then deriving, from this new ideal of theses plus hypotheses, some new constraints in terms of the free parameters ruling the geometric situation. For a toy example, consider that $x - a = 0$ is the only hypothesis, that the set of points (x, a) in this hypothesis variety is determined by the value of the parameter a, and that $x = 0$ is the (generally false) thesis. Then we add the thesis to the hypothesis, getting the new ideal $(x - a, x)$, and we observe that the elimination of x in this ideal yields the constraint $a = 0$, which is indeed the extra hypothesis we have to add to the given one $x - a = 0$, in order to have a correct statement $[x - a = 0 \wedge a = 0] \Rightarrow [x = 0]$.

With this simple idea as starting point[3], an elaborated discovery procedure, with several non trivial examples, is presented in [RV99]. It has been recently revised in [BDR] and [DR], showing that, in some precise sense, the idea of considering $H + T$ for discovering is intrinsically unique (see Section 4 for a short introduction, leading to the use of $MCCGS$ in this context).

3   Overwiew on the $MCCGS$ Algorithm

As mentioned in the introduction, specializing the basis of an ideal with parameters does not yield, in general, a basis of the specialized ideal.

This phenomenon –in the context of Gröbner basis– has been known for over fifteen years now, yielding to a rich variety of attempts towards a solution (we refer the interested reader to the bibliographic references in [MaMo06] or in [Wib06]). Finding a specializable basis (ie. providing a single basis that collects all possible bases, together with the corresponding relations among the parameters) is –more or less– the task of the different comprehensive G-Basis proposals. Although the first global solution was that of Weispfenning, as early as 1992 (see [Weis92]), the topic is quite active nowadays, as exemplified in the above quoted recent papers. The $MCCGS$ procedure, that is, computing the *minimal canonical comprehensive Gröbner system* of a given parametric ideal, is one of the approaches we are interested in. Let us describe briefly the goals and output of the $MCCGS$ algorithm.

---

[2] Already present in the well known book of [Ch88], page 72: "... The method developed here can be modified for the purpose of finding new geometry theorems... Suppose that we are trying to prove a theorem... and the final remainder... R_0 is nonzero. If we add a new hypotheses $R_0 = 0$, then we have a theorem...". Here Chou proposes adding as new hypotheses the pseudoremainder of the thesis by the ideal of hypotheses, a mathematical object which should be zero if the theorem was generally true.

[3] Indeed, things are not so trivial. Consider, for instance, $H \Rightarrow T$, where $H = (a + 1)(a + 2)(b + 1) \subset K[a, b, c]$ and $T = (a + b + 1, c) \subset K[a, b, c]$. Take as parameters $U = \{b, c\}$, a set of $\dim(H)$-variables, independent over H. Then the elimination of the remaining variables over $H + T$ yields $H' = (c, b^3 - b)$. But $H + H' = (a + 1, b, c) \cap (a + 2, b, c) \cap (a + 1, b - 1, c) \cap (a + 2, b - 1, c) \cap (b + 1, c)$ does not imply T, even if we add some non-degeneracy conditions expressed in terms of the free parameters U, since T vanishes over some components, such as $(a + 2, b - 1, c)$ (and does not vanish over some other ones, such as $(a + 1, b - 1, c)$).

Given a parametric polynomial system of equations over some computable field, such as the rational numbers, our interest focuses on discussing the type of solutions over some algebraically closed extension, such as the complex numbers, depending on the values of the parameters. Let $\mathbf{x} = (x_1, \ldots, x_n)$ be the set of variables, $\mathbf{u} = (u_1, \ldots, u_m)$ the set of parameters and $I \subset K[\mathbf{u}][\mathbf{x}]$ the parametric ideal we want to discuss, where, in order to simplify the exposition, a single field K, algebraically closed, is considered both for the coefficients and the solutions. We want to study how the solutions over K^n of the equation system defined by I vary when we specialize the values of the parameters $\mathbf{u}$ to concrete values $\mathbf{u_0} \in K$. Denote by $A = K[\mathbf{u}]$, and by $\sigma_{\mathbf{u_0}} : A[\mathbf{x}] \to K[\mathbf{x}]$ the homomorphism corresponding to the specialization (substitution of $\mathbf{u}$ by some $\mathbf{u_0} \in K$).

A Gröbner System $GS(I, \succ_{\mathbf{x}})$ of the ideal $I \subset A[\mathbf{x}]$ wrt (with respect to) the termorder $\succ_{\mathbf{x}}$ is a set of pairs (S_i, B_i), where each couple consists of a constructible set (called segment) and of a collection of polynomials, such that

$$GS(I, \succ_{\mathbf{x}}) = \{(S_i, B_i) \; : \; 1 \leq i \leq s, \; S_i \subset K^m, \; B_i \subset A[\mathbf{x}], \; \bigcup_i S_i = K^m,$$
$$\forall \mathbf{u_0} \in S_i, \; \sigma_{\mathbf{u_0}}(B_i) \text{ is a Gröbner basis of } \sigma_{\mathbf{u_0}}(I) \text{ wrt } \succ_{\mathbf{x}}\}.$$

The algorithm $MCCGS$ (Minimal Canonical Comprehensive Gröbner System) [Mo06],[MaMo06] of the ideal $I \subset A[\mathbf{x}]$ wrt the monomial order $\succ_{\mathbf{x}}$ for the variables, builds up the unique Gröbner System having the following properties:

1. The segments S_i form a partition $\mathcal{S} = \{S_1, \ldots, S_s\}$ of the parameter space K^m.
2. The polynomials in B_i are normalized to have content 1 wrt $\mathbf{x}$ over $K[\mathbf{u}]$ (in order to work with polynomials instead of with rational functions). The B_i specialize to the reduced Gröbner basis of $\sigma_{\mathbf{u_0}}(I)$, keeping the same lpp (leading power products set) for each $\mathbf{u_0} \in S_i$, i.e. the leading coefficients are different from zero on every point of S_i. [4]. Thus a concrete set of lpp can be associated to a given S_i. Often it exists a unique segment corresponding to each particular lpp, although in some cases several such segments can occur. In any case, when a segment with the reduced basis [1] exists, then it is unique. When two segments S_i, S_j share the same lpp, then there is not a common reduced basis B specializing to both B_i, B_j[5].

   Moreover, there exists a unique segment S_1 (called the *generic segment*), containing a Zariski-open set, whose associated basis B_1 is called the *generic basis* and coincides with the Gröbner basis of I considered in $K(\mathbf{u})[\mathbf{x}]$ conveniently normalized without denominators and content 1 wrt $\mathbf{x}$.
3. The partition $\mathcal{S}$ is canonical (unique for a given I and monomial order).
4. The partition is minimal, in the sense it does not exists another partition having property 2 with less sets S_i.
5. The segments S_i are described in a canonical form.

---

[4] The polynomials in the B_i's are not faithful (they do not belong to I), as they are reduced wrt to the null conditions in S_i. By abuse of language we call them *reduced bases* (i.e. not-faithful, in the terminology of Weispfenning).

[5] M. Wibmer [Wib06] has proved that for homogeneous ideals in the projective space there is at most a unique reduced basis and segment corresponding to a given lpp.

As it is known, the *lpp* of the reduced Gröbner basis of an ideal determine the cardinal or dimension of the solution set over an algebraically closed field. This makes the *MCCGS* algorithm very useful for applications as it identifies canonically the different kind of solutions for every value of the parameters. This is particularly suitable for automatic theorem proving and automatic theorem predicting, as we will show in the following sections.

Let us give an example of the output of *MCCGS*.

Example 1. Consider the system described by the following parametric ideal (here the parameters are a, b, c, d):

$$I = (x^2 + by^2 + 2cxy + 2dx, 2x + 2cy + 2d, 2by + 2cx),$$

arising in the context of finding all possible singular conics and their singularities. Calling to the Maple implementation of *MCCGS* yields a graphical and an algebraic output. The graphical output is shown in Figure 2. It contains the basic information that is to be read as follows. At the root there is the given ideal (in red). The second level (also in red) contains the *lpp* of the bases of the three different possible cases. These are [1], corresponding to the no solution (no singular points) case; [x, y], corresponding to the one solution (one singular point) case; and [x], corresponding to the case of one dimensional solution (ie. when the conic is a double line). Below each case there is a subtree (in blue) describing the corresponding S_i, with the following conventions:

- at the nodes there are ideals of $K[\mathbf{u}]$, prime in the field of definition (generated over the prime field by the coefficients of a reduced G-Basis) of the given ideal $I \subset A[\mathbf{x}]$
- a descending edge means the set theoretic "difference" of the set defined by the node above minus the set defined at the node below,
- nodes at the same level, hanging from a common node, are to be interpreted as yielding the set theoretic "union" of the corresponding sets; they form the irredundant prime decomposition of a radical ideal of $K[\mathbf{u}]$.
- every branch contains a strictly ascending chain of prime ideals.

So, in the example above, the three cases, their *lpp* and the corresponding S_i's are to be read as shown in the following table:

lpp	Basis B_i	Description of S_i
[1]	[1]	$K^3 \setminus ((\mathbb{V}(b) \setminus (\mathbb{V}(c, b) \setminus \mathbb{V}(d, c, b))) \cup \mathbb{V}(d))$
[y, x]	[2cy + d, x]	$(\mathbb{V}(b) \setminus \mathbb{V}(c, b)) \cup (\mathbb{V}(d) \setminus \mathbb{V}(d, b - c^2))$
[x]	[x + cy]	$\mathbb{V}(d, b - c^2)$

We remark that the B_i's do not appear in the Figure 2, since –in order to simplify the display– the complete bases are only given by the algebraic output of *MCCGS* and are not shown by the graphic output.

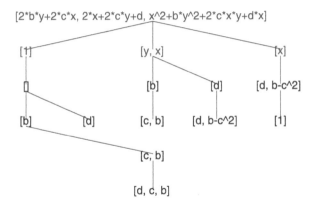

Fig. 2. *MCCGS* for the singular points of a conic

4   Using *MCCGS* for Automatic Theorem Discovering

Once we have briefly described the context for *MCCGS* and for automatic discovery, we are prepared to describe the basic idea in this paper. We can say that our goal is to show how performing a *MCCGS* procedure can improve the automatic discovery of geometry theorems.

Example 1 can be seen as a very simple example of theorem discovering. We could formulate the statement *a conic has one singular point* and try to find the conditions for the statement to be true. Without loss of generality we express the equation of the conic and its partial derivatives as

$$I = (x^2 + by^2 + 2cxy + 2dx, 2x + 2cy + 2d, 2by + 2cx),$$

and search for the values of the parameters where this system has a single solution. As shown above, we have found that the statement is true if and only if $\{b = 0, c \neq 0\}$ or if $\{d = 0, b - c^2 \neq 0\}$, since in the first segment of the table there is no solution ($B_1 = (1)$), while the third segment yields a 1-dimensional set of solutions.

In general, let $H \Rightarrow T$ be a statement expressed in terms of polynomial equations, where the ideals $H, T \subseteq K[x_1, \ldots, x_n]$ will be the corresponding hypotheses ideal and theses ideal (both, possibly, with several generators). In this context [DR] sets, as discovery goal, finding a couple of subsets of variables $U \supseteq U'$, with $X \supseteq U \supseteq U'$, and a couple of ideals $R' \subset K[U], R'' \subset K[U']$, so that the following properties hold for the associated algebraic varieties (over K^n, with K algebraically closed):

1. $\mathbb{V}(H + R'^e) \backslash \mathbb{V}(R''^e) \subseteq \mathbb{V}(T)$ (where the ^{e} stands for the extension of the ideal from its defining ring, say, $K[U]$ or $K[U']$, to $K[X]$);
2. $\mathbb{V}(H + T) \subseteq \mathbb{V}(R'^e)$;
3. $\mathbb{V}(H + R'^e) \backslash \mathbb{V}(R''^e) \neq \emptyset$.

The rationale behind such a definition is that such a couple (R', R'') is supposed to provide

- some necessary (as expressed by item 2) above)
- and sufficient (as expressed by item 1) above)

non trivial (as expressed by item 3)) complementary conditions of equality kind (given by R') and of non-degeneracy type (given by the negation of R'') for the given theses to hold under the given hypotheses.

Then it is shown in [DR] that, for a given couple of subsets of variables $U \supseteq U'$, with $X \supseteq U \supseteq U'$, there is a couple of ideals $R' \subset K[U], R'' \subset K[U']$ verifying properties 1), 2) and 3) above if and only if the couple of ideals $H' = (H + T) \cap K[U]$ and[6] $H'' = ((H + H'^e) : T^\infty) \cap K[U']$ also verify these three conditions. Moreover, Theorem 2 in [DR] shows these conditions hold if and only if $1 \notin (H')^c : H''^\infty$ (equivalently, iff $H'' \not\subseteq \sqrt{(H')^c}$), where ^{c} stands for the contraction ideal, so there is an algorithmic way of solving the posed discovery problem for a given statement and choice of variables.

Now we remark the following:

Proposition 1. *If there is a couple R', R'' verifying the above conditions, then*

$$\mathbb{V}(H + R'^e) \backslash \mathbb{V}(R''^e) = \mathbb{V}(H + H'^e) \backslash \mathbb{V}(R''^e)$$

Proof. First notice that $\mathbb{V}(H + H'^e) \subseteq \mathbb{V}(H + R'^e)$, since, by property 2), $\mathbb{V}(H + T) \subseteq \mathbb{V}(R'^e)$, thus $R'^e \subseteq \sqrt{H + T}$ and so $R' = R'^{ec} \subseteq \sqrt{H + T}^c = \sqrt{H'}$ and this implies that $\mathbb{V}(H'^e) \subseteq \mathbb{V}(R'^e)$.

Moreover, we have also that $\mathbb{V}(H + R'^e) \backslash \mathbb{V}(R''^e) \subseteq \mathbb{V}(T) \cap \mathbb{V}(H)$, by property 1), and $\mathbb{V}(T) \cap \mathbb{V}(H) \subseteq \mathbb{V}(H'^e)$, where the last inclusion follows from the definition of H'^e. We conclude that $\mathbb{V}(H + R'^e) \backslash \mathbb{V}(R''^e) \subseteq \mathbb{V}(H + H'^e) \backslash \mathbb{V}(R''^e)$.

This means that the search for candidates R' for complementary hypotheses of equality type, can be reduced to computing $\mathbb{V}(H')$. This is, precisely, the (Zariski closure of the) projection, over the parameter space of the U-variables, of $\mathbb{V}(H+T)$, and this can be computed through *MCCGS*, providing as well some other useful information (as in Corollary 1).

Proposition 2. *The projection of $\mathbb{V}(H + T)$ over the U-variables can be computed by performing a MCCGS for $I = H+T$ and $X \supset U$, discarding, if it exists, the unique segment S_i's with B_i equal to 1 and keeping the remaining S_i's.*

Proof. Since the segments of a *MCCGS* partition the parameter space U, it is enough to show that a point (u_0) is not in the projection if and only if it belongs to the S_i with associated $B_i = 1$. Now we recall that a reduced Gröbner basis is 1 if and only if the corresponding ideal is (1). Then, the ideal $H + T$ specialized at u_0 will be (1) if and only if its reduced G-basis is 1. Since we work over an algebraically closed field, this is the only case the system $H+T$, specialized at u_0,

---

[6] Let I, J be ideals of $K[X]$. Recall that $I : J = \{x, xJ \subset I\}$. Then, the *saturation of I by J* is defined as $I : J^\infty = \cup_n(I : J^n)$, cf. [KR00].

has no solution, ie. u_0 is not in the projection of $\mathbb{V}(H+T)$. But, by construction, a B_i specializes to 1 if and only if $B_i = 1$ (since the specialization must be a reduced G-basis and has the same lpp as B_i).

Corollary 1. *The union of these S_i's with associated $B_i \neq 1$ (ie. the complement of the only possible segment with $B_i = 1$) partitions the projection of $\mathbb{V}(H+T)$; that is, it holds $H \wedge T \Rightarrow \{\cup S_i\}$. Thus, the union of these S_i's provide complementary necessary conditions for the theses T to hold over H.*

We will see below (Remark 2) that, when the given statement does not hold over any geometrically meaningful component of the hypotheses variety – ie. in the automatic discovery situation– the segment with $B_i = 1$ is the generic one, so its complement provides necessary conditions for the theses T to hold over H.

Next we must study if some of these S_i's provide sufficient conditions, analyzing the behavior of each statement $H \wedge S_i \Rightarrow T$, for every segment S_i with lpp $\neq 1$. Some –perhaps all, perhaps none– of them could be true. Remark that, anyway, $H \wedge S_i \neq \emptyset$, since the associated basis in not 1. Remark, also, that *MCCGS* allows to obtain supplementary conditions S_i of the more general form (not every constructible set is the difference of two closed sets of the form $\mathbb{V}(H + R'^e) \backslash \mathbb{V}(R''^e)$, as in the previous approach).

There are some special easy cases, as shown in the next result.

Corollary 2. *For every segment S_i such that the corresponding lpp of the associated basis is, precisely, the collection of variables $\{x_1, \ldots, x_n\}$, we have that $\mathbb{V}(H) \cap S_i \subseteq \mathbb{V}(T)$, ie. $H \wedge S_i \Rightarrow T$ holds, and S_i provides sufficient conditions for T to hold over H.*

Proof. In fact, the condition on the associated *lpp* means that for every u_0 in S_i, the system $H(u_0, x) = 0, T(u_0, x) = 0$ has a unique solution, and it belongs to $\mathbb{V}(T)$. Thus $\mathbb{V}(H) \cap S_i \subseteq \mathbb{V}(T)$.

Otherwise, we should analyze, for each i with S_i involved in the projection of $\mathbb{V}(H + T)$, the validity of $H \wedge S_i \Rightarrow T$. This is a straightforward "automatic proving" step, and not of "automatic discovery", since adding again T to the collection of hypotheses $H \wedge S_i$ will not change the situation, as the projection of $\mathbb{V}(H) \cap \mathbb{V}(T) \cap S_i$ equals the projection of $\mathbb{V}(H) \cap S_i$, both being S_i.

Yet, *MCCGS* can provide a method for checking the truth of such statement $H \wedge S_i \Rightarrow T$. As it is well known, we can reformulate the hypotheses $H \wedge S_i$ as a collection of equality hypotheses H, since S_i is constructible and, then, the union of intersections of closed and open sets (in the Zariski topology). And open sets can be expressed through equalities by means of saturation techniques (such as $x \neq 0 \Leftrightarrow xy - 1 = 0$, etc.). So let us state the following propositions (adapting to the *MCCGS* context some results from [RV99], [DR]) in all generality.

Proposition 3. *Let $H \Rightarrow T$ be a statement and let U be a collection of variables independent for H. Then T vanishes identically on all the components of H where U remain independent if and only if, performing a MCCGS for $\{H, Tz - 1\}$ with respect to U, the generic basis is 1.*

Proof. Notice the stated condition on the segments of the *MCCGS* is equivalent to the fact that the contraction $(H, Tz-1) \cap K[U] \neq (0)$. In fact, this contraction is zero if and only if the projection of $\mathbb{V}(H + (Tz - 1))$ contains an open set. And this is equivalent to the fact that the generic segment has lpp $\neq 1$.

Now, if $(H, Tz - 1) \cap K[U] \neq (0)$, take some $0 \neq g \in (H, Tz - 1) \cap K[U]$. Remark that, by construction, $gT = 0$ over $\mathbb{V}(H)$. If $T \neq 0$ at some point over some component of $\mathbb{V}(H)$, then $g = 0$ over such component; so it cannot be a component where the U are independent, since $g \in K[U]$.

Conversely, if T vanishes identically over all the independent components, then we can compute an element $g \in K[U]$ vanishing over the remaining components (because U' is dependent over them). So gT vanishes all over $\mathbb{V}(H)$, and thus $(H, Tz - 1) \cap K[U'] \neq (0)$.

Remark 1. In fact, as in [CLLW], it is easy to show that the segment with associated *lpp* equal to 1 provides complementary sufficient conditions for $H \Rightarrow T$ to hold. In fact, for every $\mathbf{u_0}$ in such segment, $\mathbb{V}(H(\mathbf{u_0}, \mathbf{x}), T(\mathbf{u_0}, \mathbf{x})z - 1) = \emptyset$, so $\mathbb{V}(H(\mathbf{u_0}, \mathbf{x}) \subseteq T(\mathbf{u_0}, \mathbf{x})$. But it can happen there is no such segment.

Proposition 4. *Let $H \Rightarrow T$ be a statement and let U be a collection of variables independent for H and of dimension equal to $\dim(H)$. Then T vanishes identically on some components of H where U remains independent if and only if, performing a MCCGS for $\{H, T\}$ with respect to U, the reduced basis of the generic segment is different from 1.*

Proof. As above, the stated condition on the segments of the *MCCGS* is equivalent to the fact that the contraction $(H, T) \cap K[U'] = (0)$.

Now, if T does not vanish identically over any component of $\mathbb{V}(H)$ independent over U', the projection of $\mathbb{V}(H, T)$ over U will be a proper closed subset (since the dimension of the projection is less or equal than the dimension of the components of $\mathbb{V}(H)$ contained in $\mathbb{V}(T)$, the maximum dimension of all components of $\mathbb{V}(H)$ equals the maximum dimension of the independent components, and the dimension of the U-space equals the maximum dimension of the components of $\mathbb{V}(H)$). This contradicts the assumption $(H, T) \cap K[U] = (0)$, which implies the closure of the projection is the whole U-space.

Conversely, if T vanishes identically over some independent component (say, C) and $(H, T) \cap K[U] \neq (0)$, then we can choose an element $0 \neq g \in (H, T) \cap K[U]$. This element vanishes over any component of $\mathbb{V}(H)$ where T vanishes, in particular over C, contradicting its independence over U.

Remark 2. The last proposition can be also read in a different way: T does not vanish identically on any independent component of H if and only if the reduced basis of the generic segment is 1.

Corollary 3. *Let $H \Rightarrow T$ be a statement and let U be a collection of variables independent for H and of dimension equal to $\dim(H)$. Then T vanishes*

identically on some components of H where U remains independent and also T
does not vanish identically on some other components of H where U remains
independent if and only if

- *performing a MCCGS for $\{H, Tz - 1\}$ with respect to U, the generic segment*
  dos not have reduced basis 1, and
- *performing a MCCGS for $\{H, T\}$ with respect to U, the reduced basis of the*
  generic segment is also different from 1.

In conclusion, using $MCCGS$ one can determine, for a given statement, whether it is generally true (over all independent components, using Proposition 3, generally false (over all independent components, using Remark 2), or partially true and false (using Corollary 3). Let us call this last situation the "undecidable" case.

In fact, unfortunately, in this circumstance it is not possible, using only data on the U variables, to determine the components of H where T vanishes identically. Consider $H = b(b+1) \subseteq K[a, b]$, $T = (b)$ and take $U = \{a\}$. Here the projection of $\mathbb{V}(H, Tz - 1)$ over the U-variables is the whole a-line, so does not have any segment with lpp equal to 1, and we know the thesis does not hold over all independent components. Moreover the projection of $\mathbb{V}(H + T)$ over the U-space is again the whole a-line, so there is no segment with lpp 1, and we can conclude T holds over a component, but there is no way of separating the component $b = 0$, by manipulating H, T in terms of polynomials in the variable a.

This discussion applies to the situation described above, when considering statements $H \wedge S_i \Rightarrow T$, where segment S_i belongs to a $MCCGS$ for $\{H, T\}$ with respect to a collection U of variables and has lpp $\neq 1$. Let HH be the reformulation of $H \wedge S_i$ in terms of equalities and let (if possible) $U' \subseteq U$ be a new collection of variables, such that they are independent for HH and of dimension equal to $\dim(HH)$.

Then, as remarked above, $HH \Rightarrow T$ will be true on the segment SS_i of a $MCCGS$ with respect to $HH, Tz - 1$, with lpp 1. If it is is an open segment, then the statement $H \wedge S_i \Rightarrow T$ will be generally true (over all the components independent over U'). If it is not open segment, but there is at least one such segment, the statement will hold true under the new restrictions.

But if there is no segment at all with lpp $= 1$, then and only then we are in the undecidable case. In fact, over all points in the U'-projection of $\mathbb{V}(HH, Tz-1)$ we will have points of $\mathbb{V}(H)$ not in $\mathbb{V}(T)$ (because all the segments will have lpp $\neq 1$ in the $MCCGS$ for $HH, Tz - 1$) and also points of $\mathbb{V}(H)$ and $\mathbb{V}(T)$ (since we are also in the projection of S_i over U', and S_i corresponds to a segment of lpp $\neq 1$ for a $MCCGS$ with respect to H, T).

In this case, since the projection over U' of $\mathbb{V}(HH, T)$ will be same as the projection of $\mathbb{V}(HH)$ (both being equal to the projection of S_i), it is of no use to go further with a new discovery procedure, computing a $MCCSG$ for HH, T over U'. We know beforehand that all its segments will have lpp $\neq 1$, since over any point in the projection of S_i there will be always points on $\mathbb{V}(HH) \cap \mathbb{V}(T)$, confirming, again, that we are in the undecidable situation.

5   Examples

Let us see how this works in a collection of examples, where we have just de-tailed the discovery step (ie. computing the $MCCGS$ of $\{H, T\}$ with respect to a collection of maximal independent variables for H, and then collecting the potentially true statements $H \wedge S_i \Rightarrow T$, where segment S_i has lpp $\neq 1$) in the procedure outlined in the previous Section. That is, we have not included here the formal automatic verification in each case that the newly found hypotheses actually lead to a true statement (the "proving step").

Example 2. (See also [DR]). Next, we will develop the above introduced notions considering a statement from [Ch88] (Example 91 in his book), suitably adapted to the discovery framework. The example here is taken from [DR].

Let us consider as given data a circle and two diametral opposed points on it (say, take a circle centered at $(1, 0)$ with radius 1, and let $C = (0, 0), D = (2, 0)$ the two ends of a diameter), plus an arbitrary point $A = (u_1, u_2)$. See Figure 3. Then trace a tangent from A to the circle and let $E = (x_1, x_2)$ be the tangency point. Let $F = (x_3, x_4)$ be the intersection of DE and CA. Then we claim that $AE = AF$. Moreover, in order to be able to define the lines DE, CA, we require, as hypotheses, that $D \neq E$ (ie. $u_1 \neq 2$) and that $C \neq A$ (ie. $u_1 \neq 0$ or $u_2 \neq 0$).

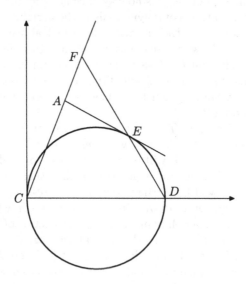

Fig. 3. Problem of Example 2

Now, using CoCoA [CNR99] and its package TP (for Theorem Proving), we translate the given situation as follows

```
Alias   TP   := $contrib/thmproving;
```

```
Use R::=Q[x[1..4],u[1..2]];
```

```
A:=[u[1],u[2]];
E:=[x[1],x[2]];
D:=[2,0];
F:=[x[3],x[4]];
C:=[0,0];


Ip1:=TP.Perpendicular([E,A],[E,[1,0]]);
Ip2:=TP.LenSquare([E,[1,0]])-1;
Ip3:=TP.Collinear([0,0],A,F);
Ip4:=TP.Collinear(D,E,F);


H:=Saturation(Ideal(Ip1,Ip2,Ip3,Ip4),Ideal(u[1]-2)*
                               Ideal(u[1], u[2]));


T:=Ideal(TP.LenSquare([A,E])-TP.LenSquare([A,F]));
```

where T is the thesis and H describes the hypothesis ideal. Notice that $Ip1$ expresses that the segments $[E, A], [E, (1,0)]$ are perpendicular; $Ip2$ states that the square of the length of $[E, (1,0)]$ is 1 (so $Ip1, Ip2$ imply E is the tangency point from A); and the next two hypotheses express that the corresponding three points are collinear. The hypothesis ideal H is here constructed by using the *saturation* command, since it is a standard way of stating that the hypothesis variety is the (Zariski) closure of the set defined by all the conditions $Ip[i] = 0, i = 1 \ldots 4$ minus the union $\{u[1] = 2\} \cup \{u[1] = 0, u[2] = 0\}$, as declared in the formulation of this example (but we refer to [DR] for a discussion on the two possible ways of introducing inequalities as hypotheses). Finally, the thesis expresses that the two segments $[AE], [AF]$ have equal non oriented length.

Now, let us use in this, clearly false, statement the approach of [RV99] or [DR] to discovery. First we check that the statement $H \rightarrow T$ is not algebraically true in any conceivable way. For instance, it turns that

```
Saturation(H, Saturation(H,T));
Ideal(1)
```
----

and this computation shows that all possible non-degeneracy conditions (those polynomials $p(\mathbf{u}, \mathbf{x})$ that could be added to the hypotheses as conditions of the kind $p(\mathbf{u}, \mathbf{x}) \neq 0$) lie in the hypotheses ideal, yielding, therefore to an empty set of conditions of the kind $p \neq 0 \land p = 0$. This implies, in particular, that the same negative result would be obtained if we restrict the computations to some subset of variables, since the thesis does not vanish on any irreducible component of the hypotheses variety.

Thus we must switch on to the discovery protocol, checking before hand that $u[1], u[2]$ actually is a (maximal) set of independent variables –the parameters– for our construction:

```
Dim(R/H);
2
```

```
Elim([x[1],x[2],x[3],x[4]],H);
Ideal(0)
```

Then we add the thesis to the hypotheses ideal and we eliminate all variables except $u[1], u[2]$

```
H':=Elim([x[1],x[2],x[3],x[4]],H+T);
H';
Ideal(-1/2u[1]^5 - 1/2u[1]^3u[2]^2 + u[1]^4)
```

```
Factor(-1/2u[1]^5 - 1/2u[1]^3u[2]^2 + u[1]^4);
[[u[1]^2 + u[2]^2 - 2u[1], 1], [u[1], 3], [-1/2, 1]]
```
----

yielding as complementary hypotheses the conditions $u[1]^2 + u[2]^2 - 2u[1] = 0 \lor u[1] = 0$ that can be interpreted by saying that either point A lies on the given circle or (when $u[1] = 0$) triangle $\Delta(A, C, D)$ is rectangle at C. In the next step of the discovery procedure we consider as new hypotheses ideal the set $H + H'$, which is of dimension 1 and where both $u[2]$ or $u[1]$ can be taken as independent variables ruling the new construction.

```
Dim(R/(H+H'));
1
```

```
Elim([x[1],x[2],x[3],x[4],u[1]],H+H');
Ideal(0)
```

```
Elim([x[1],x[2],x[3],x[4],u[2]],H+H');
Ideal(0)
```

Choosing, for example, $u[2]$ as relevant variable, we check –applying the usual automatic proving scheme– that the new statement $H \land H' \to T$ is correct under the non-degeneracy condition $u[2] \neq 0$:

```
H'':=Elim([x[1],x[2],x[3],x[4],u[1]], Saturation(H+H',T));
H'';
Ideal(u[2]^3)
```

Thus we have arrived to the following statement: Given a circle of radius 1 and centered at $(1, 0)$, and a point A not in the X-axis and lying either on the Y axis or in the circle, it holds that the segments AE, AF (where E is a tangency point from A to the circle and F is the intersection of the lines passing by $(2, 0), E$ and $A, (0, 0)$) are of equal length.

Let us now review Example 2 using *MCCGS*. As above, the hypotheses are the union of $H := H_1 \cup S$, where H_1 expresses the equality type constraints:

$$H_1 = [(x_1 - 1)(u_1 - x_1) + x_2(u_2 - x_2), (x_1 - 1)^2 + x_2^2 - 1,$$
$$u_1 x_4 - u_2 x_3, x_3 x_2 - x_4 x_1 - 2x_2 + 2x_4]$$

to which we have to add the saturation ideal expressing the inequality constraints:

$$S = [u_1x_4 - u_2x_3, x_1u_1 - u_1 - x_1 + x_2u_2, x_4x_2 - 2x_2u_2 - x_3u_1 + 2u_1,$$
$$x_4x_1 - 2x_1u_2 + u_2x_3, x_3x_2 - 2x_1u_2 + u_2x_3 - 2x_2 + 2x_4,$$
$$x_1x_3 + x_3u_1 + 2x_2u_2 - 2x_1 - 2u_1, x_1^2 - 2x_1 + x_2^2,$$
$$x_3u_1^2 + 2x_2u_2u_1 - 2u_2^2x_1 + u_2^2x_3 - 2u_1^2 - 2x_2u_2 + 2u_2x_4,$$
$$x_3^2u_1 + x_4u_2x_3 + 2x_4^2 - 4x_3u_1 - 4u_2x_4 + 4u_1,$$
$$u_1x_2^2 - x_1x_2u_2 - x_2^2 + x_2u_2 + x_1 - u_1,$$
$$u_2x_3^3 + u_2x_4^2x_3 + 2x_4^3 - 4u_2x_3^2 - 4u_2x_4^2 + 4u_2x_3].$$

The thesis is

$$T = (u_1 - x_1)^2 + (u_2 - x_2)^2 - (u_1 - x_3)^2 - (u_2 - x_4)^2.$$

Calling now mccgs($H_1 \cup S \cup T$, lex(x_1, x_2, x_3, x_4), lex(u_1, u_2)) one obtains the following segments:

Segment	lpp	Description of S_i
1	[1]	$K^2 \setminus (\mathbb{V}(u_1^2 + u_2^2 - 2u_1) \cup \mathbb{V}(u_1))$
2	$[x_4^2, x_3, x_2, x_1]$	$\mathbb{V}(u_1^2 + u_2^2 - 2u_1) \setminus (\mathbb{V}(u_1 - 2, u_2) \cup \mathbb{V}(u_1, u_2))$
3	$[x_4^2, x_3, x_2, x_1]$	$\mathbb{V}(u_1) \setminus (\mathbb{V}(u_1, u_2^2 + 1) \cup \mathbb{V}(u_1, u_2))$
4	$[x_4, x_3, x_2, x_1]$	$\mathbb{V}(u_1, u_2^2 + 1)$
5	$[x_4^2, x_3, x_2^2, x_1]$	$\mathbb{V}(u_1 - 2, u_2)$
6	$[x_4^2, x_3^2, x_2, x_1]$	$\mathbb{V}(u_2, u_1)$

Segment S_1 states that point $A(u_1, u_2)$ must lie either in the Y-axis or on the circle, as a necessary condition in the parameter space $\mathbf{u} = (u_1, u_2)$ for the existence of solutions, in the hypotheses plus thesis variety, lying over $\mathbf{u}$. This essentially agrees with the result obtained in [DR].

A detailed analysis of the remaining segments show a variety of formulas for determining the (sometimes not unique) values of points $E(x_1, x_2)$ and $F(x_3, x_4)$ –verifying the theorem– over the corresponding parameter values.

For completeness we give the different bases associated, in the different segments, to the above ideal of thesis plus hypotheses

$B_1 = [1]$
$B_2 = [u_2^2 + x_4^2 - 2u_2x_4, -u_1x_4 + u_2x_3, u_2^3 - 2u_2u_1 + x_2u_2^2 + (-2u_2^2 + 2u_1)x_4,$
$\quad\quad u_2u_1 + x_1u_2 - 2u_1x_4]$
$B_3 = [-2u_2x_4 + x_4^2, x_3, (u_2^2 + 1)x_2 - x_4, (u_2^2 + 1)x_1 - u_2x_4]$
$B_4 = [x_4, x_3, x_2, x_1]$
$B_5 = [x_4, -4 + 2x_3, x_2^2, -2 + x_1]$
$B_6 = [x_4^2, x_3^2, -x_3x_4 + 2x_2 - 2x_4, 2x_1]$

Example 3. Next we consider the problem[7] described in Figure 4. Take a circle C with center at $O(0,0)$ and radius 1 and let us denote points $A = (-1, 0)$ and

---

[7] We thankfully acknowledge here that this problem was suggested by a colleague, Manel Udina.

$B = (0,1)$. Let D be an arbitrary point with coordinates $D = (1+a,b)$ and let $C = (1+a,0)$ be another point in the X-axis, lying under point D. Then trace the line BC. Assume this line intersects the circle $\mathcal{C}$ at point $P(x,y)$.

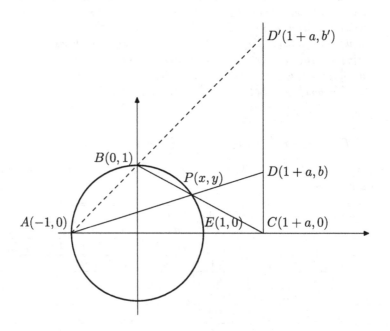

Fig. 4. Example 3

Consider now the, in general false, statement *"the points A, P, D are aligned"*. We want to discover the conditions on the parameters a, b for the statement to be true. The set of hypotheses plus thesis equations are very simple:

$$HT = [x^2 + y^2 - 1, -x + 1 - y + a - ay, -2y + b + xb - ay]$$

Take x, y as variables and a, b as parameters and call mccgs(HT, lex(x,y), lex(a,b)). The graphical output of the algorithm can be seen in Figure 5, and the algebraic description appears in the following table.

lpp	Basis B_i	Description of S_i
[1]	[1]	$(K^2 \setminus (V(a-b) \setminus V(a-b,(b+1)^2+1)))$ $\cup (K^2 \setminus V(2+a))$ $\cup (K^2 \setminus V(a-b+2))$
$[y,x]$	$[(ab+a+b+2)y - 2b - ab,$ $(ab+a+b+2)x + b + ab - 2 - 3a - a^2]$	$(V(a-b) \setminus V(a-b,(b+1)^2+1))$ $\cup (V(2+a) \setminus V(b,2+a))$ $\cup (V(a-b+2) \setminus V(b,2+a))$
$[y^2,x]$	$[y(y-1), 1+x-y]$	$V(b,2+a)$

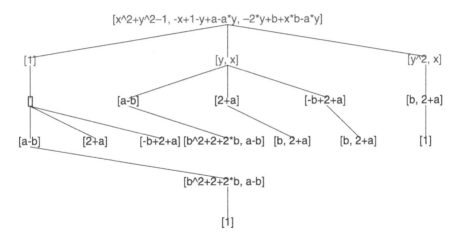

Fig. 5. Canonical tree for Example 3

As we see, the generic case has basis [1] showing that the statement is false in general. The interesting case corresponds, as it is usually expected, to the case with lpp $= [x, y]$, providing a unique solution for P. The description of the parameter set associated to this basis gives the union of three different locally closed sets, namely $\mathbb{V}(a - b) \setminus \mathbb{V}(a - b, (b + 1)^2 + 1)$, $\mathbb{V}(2 + a) \setminus \mathbb{V}(b, 2 + a)$ and $\mathbb{V}(a - b + 2) \setminus \mathbb{V}(b, 2 + a)$, expressing complementary hypotheses for the statement to hold.

The first set is (perhaps) the expected one, corresponding to the case $a = b$ (except for the degenerate complex point (b, b) with $(b + 1)^2 + 1 = 0$, without interest from the real point of view). Thus we can say that the statement holds if point C is equidistant from point D and point E.

The second set yields $a = -2$ and corresponds to the situation where point D is on the tangent to the circle trough the point $(-1, 0)$ (except for the degenerate case $b = 0$). In this case $P = A$ and, obviously, A, P, D are aligned (even in the degenerate case, as stated in the third segment, corresponding to the lpp $= [y^2, x]$).

Finally, the third set gives the condition $b = a + 2$ and it is also interesting, since it corresponds to the case where the intersecting point of the line BC with the circle is taken to be B instead of P, and then point D' should be in the vertical of C and at distance $D'C$ equal to distance EC plus two.

Example 4. [Isosceles orthic triangle]

In [DR] the conditions for the orthic triangle of a given triangle (that is, the triangle built up by the feet of the altitudes of the given triangle over each side) to the equilateral have been discovered. Next example aims to discover conditions for a given triangle in order to have an isosceles orthic triangle.

Consider the triangle of Figure 1 with vertices $A(-1, 0)$, $B(1, 0)$ and $C(a, b)$, corresponding to a generic triangle having one side of length 2. Denote by

$P_1(a, 0)$, $P_2(x_2, y_2)$, $P_3(x_3, y_3)$ the feet of the altitudes of the given triangle, ie. the vertices of the orthic triangle. The equations defining these vertices are:

$$H = \left.\begin{array}{r} (a-1)\, y_2 - b\,(x_2 - 1) = 0, \\ (a-1)\,(x_2 + 1) + b\, y_2 = 0, \\ (a+1)\, y_3 - b\,(x_3 + 1) = 0, \\ (a+1)\,(x_3 - 1) + b\, y_3 = 0, \end{array}\right\}$$

Now let us add the condition $\overline{P_1 P_3} = \overline{P_1 P_2}$.

$$T = (x_3 - a)^2 + y_3^2 - (x_2 - a)^2 - y_2^2 = 0.$$

Take x_2, x_3, y_2, y_3 as variables and a, b as free parameters and call

$$\mathrm{mccgs}(H \cup T, \mathrm{lex}(x_2, x_3, y_2, y_3), \mathrm{lex}(a, b)).$$

The output has now four segments. The generic case, with lpp $= [1]$, meaning that the orthic triangle is, in general, not isosceles; one interesting case with lpp $= [y_3, y_2, x_3, x_2]$; and two more cases we can call degenerate, with *lpp*'s $[y_2, x_3^2, x_2]$ and $[y_2, x_3, x_2^2]$, respectively. For the interesting case we show the graphic output in Figure 6. Its basis is

$$B_2 = [(a^2 + b^2 + 2a + 1)y_3 - 2ab - 2b, (a^2 + b^2 - 2a + 1)y_2 + 2ab - 2b,$$
$$(a^2 + b^2 + 2a + 1)x_3 - a^2 + b^2 - 2a - 1, (a^2 + b^2 - 2a + 1)x_2 + a^2 - b^2 - 2a + 1].$$

Next table shows the description of the *lpp* and the S_i's for the the four cases:

lpp	Description of S_i
$[1]$	$K^2 \setminus ((V(a) \setminus V(b^2 + 1, a))$
	$\cup\ (V(a^2 - b^2 - 1) \setminus V(b^2 + 1, a))$
	$\cup\ V(a^2 + b^2 - 1))$
$[y_3, y_2, x_3, x_2]$	$V(a) \setminus V(b^2 + 1, a)$
	$\cup\ V(a^2 + b^2 - 1) \setminus (V(b, a - 1) \cup V(b, a + 1))$
	$\cup\ (V(a^2 - b^2 - 1) \setminus (V(b^2 + 1, a) \cup V(b, a - 1) \cup V(b, a + 1)))$
$[y_2, x_3^2, x_2]$	$V(b, a + 1)$
$[y_2, x_3, x_2^2]$	$V(b, a - 1)$

The description of the parameter set (over the reals) for which the theorem is potentially true and no degenerate can be phrased as follows:

1) $a = 0$
2) $a^2 + b^2 = 1$ except the points $(1, 0)$ and $(-1, 0)$
3) $a^2 - b^2 = 1$ except the points $(1, 0)$ and $(-1, 0)$

This set is represented in Figure 7. and corresponds to

1) The given triangle is itself isosceles ($a = 0$);
2) The given triangle is rectangular at vertex C (with vertices $A(-1, 0)$, $B(1, 0)$ and the vertex $C(a, b)$ inscribed in the circle $a^2 + b^2 = 1$,

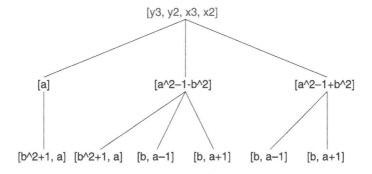

Fig. 6. Canonical tree branch for lpp $= [y_3, y_2, x_3, x_2]$ in Example 4

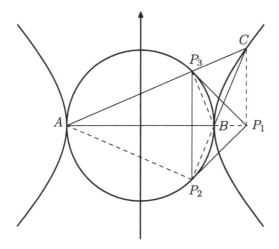

Fig. 7. Solutions of Example 4

3) The given triangle has vertices $A(-1, 0)$, $B(1, 0)$ and vertex $C(a, b)$ lies on the hyperbola $a^2 - b^2 = 1$.

Solution 1) is, perhaps, not surprising. Solution 2) corresponds to rectangular triangles for which the orthic triangle reduces to a line, that can be considered a degenerate isosceles triangle. But solution 3) is a nice novelty: it exists a one parameter family of non-isosceles triangles having isosceles orthic triangles.

The remaining two cases in the $MCCGS$ output with lpp $= [y_2, x_3^2, x_2]$ and lpp$= [y_2, x_3, x_2^2]$ represent degenerate triangles without geometric interest (namely $C = A$ and $C = B$).

Thus, after performing an automatic proving procedure for the new hypotheses, we can formulate the following theorem:

Theorem 1. *Given a triangle with vertices $A(-1, 0)$, $B(1, 0)$ and $C(a, b)$, its orthic triangle will be isosceles if and only if vertex C lies either on the line*

$a = 0$ *(and then the given triangle is itself isosceles) or in the circle* $a^2 + b^2 = 1$ *(and then it is rectangular) or in the hyperbola* $a^2 - b^2 = 1$.

Example 5. [Skaters]

Our final example is taken from the pastimes section of the French journal *Le Monde*, published on the printed edition of Jan. 8, 2007. This example is there attributed to E. Busser and G. Cohen. We think it is nice from *Le Monde* to include the proof of a theorem as a pastime. Actually, the statement to be proved was presented as arising from a more down-to-earth situation: two ice-skaters are moving forming two intersecting circles, at same speed and with the same sense of rotation. They both depart from one of the points of intersection of the two circles. Then the journal asked to show that the two skaters were always aligned with the other point of intersection (where some young lady, both skaters were interested at, was placed...).

Let us translate this problem into a theorem discovering question, as follows.

We will consider two circles with centers at $P(a, 1)$ and $Q(-b, 1)$ and radius $r_1^2 = a^2 + 1$ and $r_2^2 = b^2 + 1$, as shown in Figure 8, intersecting at points $O(0, 0)$ and $M(0, 2)$. Consider generic points –the skaters– $A(x_1, y_1)$ and $B(x_2, y_2)$ on the respective circles. Point A will be parametrized by the oriented angle $v = \widehat{OPA}$ and, correspondingly, point B will describe the oriented angle $w = \widehat{OQB}$. Therefore we can say that angle zero corresponds to the departing location of both skaters, namely, point O.

We claim that, for whatever position of points A, B, *the points* A, M, B *are aligned*, which is obviously false in general. But we want to determine if there is a relation between the two oriented angles making this statement to hold true. Denote c_v, s_v, c_w, s_w the cosine and sine of the angles v and w. It is easy to establish the basic hypotheses, using scalar products:

$$H_1 = [(x_1 - a)^2 + (y_1 - 1)^2 - a^2 - 1, (x_2 + b)^2 + (y_2 - 1)^2 - b^2 - 1,$$
$$a(x_1 - a) + (y_1 - 1) + (a^2 + 1)c_v, -b(x_2 + b) + (y_2 - 1) + (1 + b^2)c_w$$

Now, as the angles are to be taken oriented (because we assume the skaters tare moving on the corresponding circle in the same sense), we need to add the vectorial products involving also the sine to determine exactly the angles and not only their cosines. So we add the hypotheses:

$$H_2 = [a(y_1 - 1) - (x_1 - a) + (a^2 + 1)s_v, -b(y_2 - 1) - (x_2 + b) + (b^2 + 1)s_w]$$

The thesis is, clearly:

$$T = x_1 y_2 - 2x_1 - x_2 y_1 + 2x_2.$$

The radii of the circles are

$$r_1^2 = a^2 + 1 \text{ and } r_2^2 = b^2 + 1$$

and for $r_1 \neq 0$ and $r_2 \neq 0$ we have

$$c_{v_0} = \cos v_0 = \cos \widehat{OPM} = \frac{a^2 - 1}{a^2 + 1}, \ s_{v_0} = \sin v_0 = \sin \widehat{OPM} = \frac{-2a}{a^2 + 1},$$
$$c_{w_0} = \cos w_0 = \cos \widehat{OQM} = \frac{b^2 - 1}{b^2 + 1}, \ s_{w_0} = \sin w_0 = \sin \widehat{OQM} = \frac{2b}{b^2 + 1}.$$

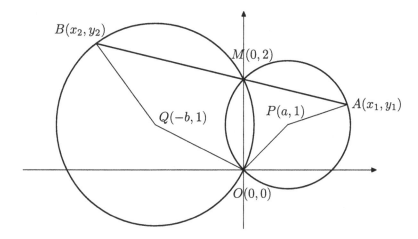

Fig. 8. Skaters problem

We want to take a, b and the angles v and w –in terms of the sines and cosines–
as parameters. So we must introduce the constraints on the sine and cosine parameters. Moreover, we notice there are also some obvious degenerate situations,
namely $r_1 = 0$, $r_2 = 0$ and $a + b = 0$, corresponding to null radii or coincident
circles, and we want to avoid them.

Currently, *MCCGS* allows us to introduce all these constraints in order to
discuss the parametric system. The call is now

$$\text{mccgs}(H_1 \cup H_2 \cup T, \ \text{lex}(x_1, y_1, x_2, y_2), \ \text{lex}(a, b, s_v, c_v, s_w, c_w),$$
$$\text{null} = [c_v^2 + s_v^2 - 1, c_w^2 + s_w^2 - 1], \ \text{notnull} = \{a^2 + 1, b^2 + 1, a + b\}).$$

including the constraints on the parameters and eluding degenerate situations
as options for *MCCGS*.

The result is that *MCCGS* outputs only 2 cases. The first one has basis [1],
showing that, in general, there is no solution to our query. The second one has
lpp $= [y_2, x_2, y_1, x_1]$ determining in a unique form the points A and B for the
given values of the parameters. The associated basis is

$$[y_2 + c_w - bs_w - 1, x_2 - bc_w - s_w + b, y_1 + c_v + as_v - 1, x_1 + ac_v - s_v - a]$$

with parameter conditions that can be expressed as the union of three irreducible
varieties:

$$V_1 = \mathbb{V}(c_w^2 + s_w^2 - 1, c_v - c_w, s_v - s_w)$$
$$V_2 = \mathbb{V}(c_w^2 + s_w^2 - 1, c_v^2 + s_v^2 - 1, s_w + bc_w - b, bs_w - c_w - 1)$$
$$V_3 = \mathbb{V}(c_w^2 + s_w^2 - 1, c_v^2 + s_v^2 - 1, -s_v + ac_v - a, as_v + c_v + 1)$$

The interpretation is easy: V_1 corresponds to arbitrary a, b, w, plus the essential condition $v = w$, which is the interesting case, stating that our conjecture

requires (and it is easy to show that this condition is sufficient) that both skaters keep moving with the same angular speed.

V_2 corresponds to $s_w = s_{w_0}$, $c_w = c_{w_0}$ and a, b, v free, thus $B = M$ and A can take any position.

V_3 is analogous to V_2, and corresponds to placing $A = M$ and B anywhere.

So we can summarize the above discussion in the following

Theorem 2. *Given two non coincident circles of non-null radii and centers P and Q, intersecting at two points O and M, let us consider points A, B on each of the circles. Then the three points A, M, B are aligned if and only if the oriented angles $\widehat{OPA}$ and $\widehat{OQB}$ are equal or A or B or both coincide with M.*

6   Performances

Although the principal advantage of *MCCGS* in relation to other CGS algorithms is the simplicity and properties of the output: the minimal number of segments and the characterization of the type of the solution depending on the values of the parameters, the computer implementation[8] of the corresponding package, named *dpgb* release 7.0, in *Maple 8* is relatively short time consuming. Moreover,we think that no other actual PCAD software will be able to obtain the accurate result obtained, for example, in example 13. We give here a table with the CPU time and number of segments for the examples of the paper.

Example	CPU time (sec.)	Number of segments
1	1.9	3
2	12.8	6
3	0.98	3
4	4.4	4
5	129.4	2

The computations were done with a Pentium(R) 4 CPU at 3.40 Ghz and 1.00 GB RAM.

7   Conclusion

We have briefly introduced the principles of automatic discovery and also the ideas –in the context of comprehensive Gröbner basis– for discussing polynomial systems with parameters, via the new *MCCGS* algorithm. Then we have shown how natural is to merge both concepts, since the parameter discussion can be interpreted as yielding, in particular, the projection of the system solution set over the parameter space; and since the conditions for discovery can be obtained by the elimination of the dependent variables over the ideal of hypotheses and

---

[8] That can be freely obtained at http://www-ma2.upc.edu/~montes

thesis. Moreover, we have also remarked how the approach through *MCCGS* provides new candidate complementary conditions of more general type and, in some particular instances (segments of the parameter space yielding to unique solution), quite common in our examples, an easy test for the sufficiency of these conditions. Finally, the use of *MCCGS* for automatic proving has been presented, as part of a formal discussion on the limitations of the discovery method.

We have exemplified this approach through a collection of non-trivial examples (performed by running the current Maple implementation of *MCCGS*, see [MaMo06], over a laptop, without special time – a few seconds– or memory requirements), showing that in all cases, the *MCCGS* output is very suitable to providing geometric insight, allowing the actual discovery of interesting and new? theorems (and pastimes!).

References

[BDR]   Beltrán, C., Dalzotto, G., Recio, T.: The moment of truth in automatic theorem proving in elementary geometry. In: Botana, F., Roanes-Lozano, E. (eds.) Proceedings ADG 2006 (extended abstracts), Universidad de Vigo (2006)

[BR05]   Botana, F., Recio, T.: Towards solving the dynamic geometry bottleneck via a symbolic approach. In: Hong, H., Wang, D. (eds.) ADG 2004. LNCS (LNAI), vol. 3763, pp. 92–111. Springer, Heidelberg (2006)

[CLLW]   Chen, X.F., Li, P., Lin, L., Wang, D.K.: Proving geometric theorems by partitioned-parametric Gröbner bases. In: Hong, H., Wang, D. (eds.) ADG 2004. LNCS (LNAI), vol. 3763, pp. 34–44. Springer, Heidelberg (2005)

[CW]   Chen, X.F., Wang, D.K.: The projection of a quasivariety and its application on geometry theorem proving and formula deduction. In: Winkler, F. (ed.) ADG 2002. LNCS (LNAI), vol. 2930, pp. 21–30. Springer, Heidelberg (2004)

[Ch84]   Chou, S.-C.: Proving Elementary Geometry Theorems Using Wu's Algorithm Contemporary Mathematics. Automated Theorem Proving: After 25 Years, American Mathematical Society, Providence, Rhode Island 29, 243–286 (1984)

[Ch85]   Chou, S.-C.: Proving and discovering theorems in elementary geometries using Wu's method Ph.D. Thesis, Department of Mathematics, University of Texas, Austin (1985)

[Ch87]   Chou, S.C.: A Method for Mechanical Derivation of Formulas in Elementary Geometry. Journal of Automated Reasoning 3, 291–299 (1987)

[Ch88]   Chou, S.-C.: Mechanical Geometry Theorem Proving. Mathematics and its Applications, D. Reidel Publ. Comp. (1988)

[ChG90]   Chou, S.-C., Gao, X.-S.: Methods for Mechanical Geometry Formula Deriving. In: Proceedings of International Symposium on Symbolic and Algebraic Computation, pp. 265–270. ACM Press, New York (1990)

[CNR99]   Capani, A., Niesi, G., Robbiano, L.: CoCoA, a System for Doing Computations in Commutative Algebra. The version 4.6 is available at the web site, http://cocoa.dima.unige.it

[DR]   Dalzotto, G., Recio, T.: On protocols for the automated discovery of theorems in elementary geometry. J. Automated Reasoning (submitted)

[DG]   Dolzmann, A., Gilch, L.: Generic Hermitian Quantifier Elimination. In: Buchberger, B., Campbell, J.A. (eds.) AISC 2004. LNCS (LNAI), vol. 3249, pp. 80–93. Springer, Heidelberg (2004)

[DSW]   Dolzmann, A., Sturm, T., Weispfenning, V.: A new approach for automatic theorem proving in real geometry. Journal of Automated Reasoning 21(3), 357–380 (1998)

[K86]   Kapur, D.: Using Gröbner basis to reason about geometry problems. J. Symbolic Computation 2(4), 399–408 (1986)

[K89]   Kapur, D.: Wu's method and its application to perspective viewing. In: Kapur, D., Mundy, J.L. (eds.) Geometric Reasoning, The MIT press, Cambridge (1989)

[K95]   Kapur, D.: An approach to solving systems of parametric polynomial equations. In: Saraswat, Van Hentenryck (eds.) Principles and Practice of Constraint Programming, MIT Press, Cambridge (1995)

[KR00]   Kreuzer, M., Robbiano, L.: Computational Commutative Algebra 1. Springer, Heidelberg (2000)

[Ko]   Koepf, W.: Gröbner bases and triangles. The International Journal of Computer Algebra in Mathematics Education 4(4), 371–386 (1998)

[MaMo05]   Manubens, M., Montes, A.: Improving DISPGB Algorithm Using the Discriminant Ideal. Jour. Symb. Comp. 41, 1245–1263 (2006)

[MaMo06]   Manubens, M., Montes, A.: Minimal Canonical Comprehensive Groebner System. arXiv: math.AC/0611948. (2006)

[Mo02]   Montes, A.: New Algorithm for Discussing Gröbner Bases with Parameters. Jour. Symb. Comp. 33(1-2), 183–208 (2002)

[Mo06]   Montes, A.: About the canonical discussion of polynomial systems with parameters. arXiv: math.AC/0601674 (2006)

[R98]   Recio, T.: Cálculo simbólico y geométrico. Editorial Síntesis, Madrid (1998)

[RV99]   Recio, T., Pilar, M., Pilar Vélez, M.: Automatic Discovery of Theorems in Elementary Geometry. J. Automat. Reason. 23, 63–82 (1999)

[RB]   Recio, T., Botana, F.: Where the truth lies (in automatic theorem proving in elementary geometry). In: Laganà, A., Gavrilova, M., Kumar, V., Mun, Y., Tan, C.J.K., Gervasi, O. (eds.) ICCSA 2004. LNCS, vol. 3044, pp. 761–771. Springer, Heidelberg (2004)

[Ri]   Richard, P.: Raisonnement et stratégies de preuve dans l'enseignement des mathématiques. Peter Lang Editorial, Berne (2004)

[SS]   Seidl, A., Sturm, T.: A generic projection operator for partial cylindrical algebraic decomposition. In: Sendra, R. (ed.) ISSAC 2003. Proceedings of the 2003 International Symposium on Symbolic and Algebraic Computation, Philadelphia, Pennsylvania, pp. 240–247. ACM Press, New York (2003)

[St]   Sturm, T.: Real Quantifier Elimination in Geometry. Doctoral dissertation, Department of Mathematics and Computer Science. University of Passau, Germany, D-94030 Passau, Germany (December 1999)

[Wa98]   Wang, D.: Gröbner Bases Applied to Geometric Theorem Proving and Discovering. In: Buchberger, B., Winkler, F. (eds.) Gröbner Bases and Applications, 251th edn. London Mathematical Society Lecture Notes Series, vol. 251, pp. 281–301. Cambridge University Press, Cambridge (1998)

[Weis92]   Weispfenning, V.: Comprehensive Grobner bases. Journal of Symbolic Computation 14(1), 1–29 (1992)

[Wib06]   Wibmer, M.: Gröbner bases for families of affine or projective schemes. arXiv: math.AC/0608019. (2006)

Mechanical Theorem Proving in Tarski's Geometry

Julien Narboux

Équipe LogiCal, LIX
École Polytechnique, 91128 Palaiseau Cedex, France
Julien.Narboux@inria.fr
http://www.lix.polytechnique.fr/Labo/Julien.Narboux/

Abstract. This paper describes the mechanization of the proofs of the first height chapters of Schwabäuser, Szmielew and Tarski's book: *Metamathematische Methoden in der Geometrie*. The proofs are checked formally using the Coq proof assistant. The goal of this development is to provide foundations for other formalizations of geometry and implementations of decision procedures. We compare the mechanized proofs with the informal proofs. We also compare this piece of formalization with the previous work done about Hilbert's *Grundlagen der Geometrie*. We analyze the differences between the two axiom systems from the formalization point of view.

1 Introduction

Euclid is considered as the pioneer of the axiomatic method, in the *Elements*, starting from a small number of self-evident truths, called postulates or common notions, he derives by purely logical rules most of the geometrical facts that were discovered in the two or three centuries before him. But upon a closer reading of Euclid's *Elements*, we find that he does not adhere as strictly as he should to the axiomatic method. Indeed, at some steps in certain proofs he uses a method of "superposition of triangles". This kind of justifications can not be derived from his set of postulates.

In 1899, in *der Grundlagen der Geometrie*, Hilbert described a more formal approach and proposed a new axiom system to fill the gaps in Euclid's system.

Recently, the task consisting in mechanizing Hilbert's *Grundlagen der Geometrie* has been partially achieved. A first formalization using the Coq proof assistant [1] was proposed by Christophe Dehlinger, Jean-François Dufourd and Pascal Schreck [2]. This first approach was realized in an intuitionist setting, and concluded that the decidability of point equality and collinearity is necessary to check Hilbert's proofs. Another formalization using the Isabelle/Isar proof assistant [3] was performed by Jacques Fleuriot and Laura Meikle [4]. Both formalizations have concluded that, even if Hilbert has done some pioneering work about formal systems, his proofs are in fact not fully formal, in particular degenerated cases are often implicit in the presentation of Hilbert. The proofs can be made more rigorous by machine assistance. Indeed, in the different editions

F. Botana and T. Recio (Eds.): ADG 2006, LNAI 4869, pp. 139–156, 2007.

of *die Grundlagen der Geometrie* the axioms were changed, but the proofs were note always changed accordingly, this obviously resulted in some inconsistencies. The use of a proof assistant solves this problem: when an axiom is changed it is easy to check if the proofs are still valid.

In the early 60s, Wanda Szmielew and Alfred Tarski started the project of a treaty about the foundations of geometry based on another axiom system for geometry designed by Tarski in the 20s[1]. A systematic development of euclidean geometry was supposed to constitute the first part but the early death of Wanda Szmielew put an end to this project. Finally, Wolfram Schwabhäuser continued the project of Wanda Szmielew and Alfred Tarski. He published the treaty in 1983 in German: *Metamathematische Methoden in der Geometrie* [6]. In [7], Art Quaife used a general purpose theorem prover to automate the proof of some lemmas in Tarski's geometry.

In this paper we describe our formalization of the first eight chapters of the book of Wolfram Schwabhäuser, Wanda Szmielew and Alfred Tarski in the Coq proof assistant.

We will first describe the different axioms of Tarski's geometry and give an history of the different versions of this axiom system. Then after a shot introduction to the system Coq, we present our formalization of the axiom system and the mechanization of one example theorem. Finally, we compare our formalization with existing ones and compare Tarski's axiomatic system with Hilbert's system from the mechanization point of view.

2 Motivations

We aim at two applications: the first one is the use of a proof assistant in the education to teach geometry [8,9], the second one is the proof of programs in the field computational geometry.

These two themes have already been partially addressed by the community. Frédérique Guilhot has realized a large Coq development about euclidean geometry following a presentation suitable for use in french high-school [10]. Concerning the proof of programs in the field of computational geometry we can cite the formalization of convex hulls algorithms by David Pichardie and Yves Bertot in Coq [11] and by Laura Meikle and Jacques Fleuriot in Isabelle [12] and the formalization of an image segmentation algorithm by Jean-François Dufourd [13]. In [14,15], we have presented the formalization and implementation in the Coq proof assistant of the area decision procedure of Chou, Gao and Zhang [16].

Formalizing geometry in a proof assistant has not only the advantage of providing a very high level of confidence in the proof generated, it also permits to insert purely geometric arguments within other kind of proofs such as for instance proof of correctness of programs or proofs by induction. For the time being all the formal developments we have cited are distinct and as they do not use the same axiomatic system, they can not be combined.

---

[1] These historical pieces of information are taken from the introduction of the publication by Givant in 1999 [5] of a letter from Tarski to Schwabhäuser (1978).

The goal of our mechanization is to do a first step in the direction of the merging of these developments. We aim at providing very clear foundations for other formalizations of geometry and implementations of decision procedures.

3   Tarski's Axiom System

Alfred Tarski worked on the axiomatization and meta-mathematics of euclidean geometry from 1926 until his death in 1983. Several axiom systems were produced by Tarski and his students. In this section, we first give an informal description of the propositions which appeared in the different versions of Tarski's axiom system, then we provide an history of these versions and finally we present the version we have formalized.

The axioms are based on first order logic and two predicates:

betweenness. The ternary *betweenness* predicate $\beta\ A B C$ informally states that B lies on the line AC between A and C.

equidistance. The quaternary *equidistance* predicate $AB \equiv CD$ informally means that the distance from A to B is equal to the distance from C to D.

Note that in Tarski's geometry, only a set of points is assumed, in particular, lines are *defined* by two distinct points whereas in Hilbert's axiom system lines and planes are *assumed*.

3.1   Axioms

We reproduce here the list of propositions which appear in the different versions of Tarski's axiom system. We adopt the same numbering as in [5]. Free variables are considered to be implicitly quantified universally.

1 Symmetry for equidistance

$$AB \equiv BA$$

2 Pseudo-transitivity for equidistance[2]

$$AB \equiv PQ \wedge AB \equiv RS \Rightarrow PQ \equiv RS$$

3 Identity for equidistance

$$AB \equiv CC \Rightarrow A = B$$

4 Segment construction

$$\exists X, \beta\ Q\ A\ X \wedge AX \equiv BC$$

---

[2] Note that we call this property *pseudo*-transitivity because the transitivity property for equidistance should be:

$$AB \equiv PQ \wedge PQ \equiv RS \Rightarrow AB \equiv RS.$$

The segment construction axiom states that one can build a point on a ray at a given distance.

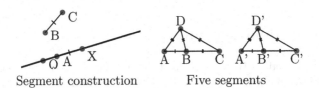

Segment construction   Five segments

5 Five segments

$$A \neq B \wedge \beta\,ABC \wedge \beta\,A'B'C' \wedge$$
$$AB \equiv A'B' \wedge BC \equiv B'C' \wedge AD \equiv A'D' \wedge BD \equiv B'D' \Rightarrow CD \equiv C'D'$$

5_1 Five segments (variant)

$$A \neq B \wedge B \neq C \wedge \beta\,ABC \wedge \beta\,A'B'C' \wedge$$
$$AB \equiv A'B' \wedge BC \equiv B'C' \wedge AD \equiv A'D' \wedge BD \equiv B'D' \Rightarrow CD \equiv C'D'$$

This second version differs from the first one only by the condition $B \neq C$.

6 Identity for betweenness

$$\beta\,ABA \Rightarrow A = B$$

The original Pasch axiom states that if a line intersects one side of a triangle and misses the three vertexes, then it must intersect one of the other two sides.

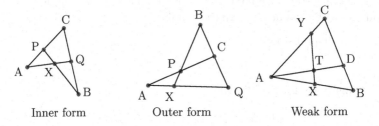

Inner form   Outer form   Weak form

Fig. 1. Axioms of Pasch

7 Pasch (inner form)

$$\beta\,APC \wedge \beta\,BQC \Rightarrow \exists X, \beta\,PXB \wedge \beta\,QXA$$

7_1 Pasch (outer form)

$$\beta\,APC \wedge \beta\,QCB \Rightarrow \exists X, \beta\,AXQ \wedge \beta\,BPX$$

7_2 Pasch (outer form) (variant)

$$\beta\,APC \wedge \beta\,QCB \Rightarrow \exists X, \beta\,AXQ \wedge \beta\,XPB$$

7_3 weak Pasch

$$\beta\,AT\,D \wedge \beta\,B\,D\,C \Rightarrow \exists X, Y, \beta\,A\,X\,B \wedge \beta\,A\,Y\,C \wedge \beta\,Y\,T\,X$$

Dimension axioms provide upper and lower bound for the dimension of the space. Note that lower bound axioms for dimension n are the negation of upper bound axioms for the dimension $n-1$.

8(2) Dimension, lower bound 2

$$\exists ABC, \neg\beta\,A\,B\,C \wedge \neg\beta\,B\,C\,A \wedge \neg\beta\,C\,A\,B$$

There are three non collinear points.

8(n) Dimension, upper bound n

$$\bigwedge_{1 \le i < j < n} P_i \neq P_j \wedge$$
$$\exists ABCP_1 P_2 \ldots P_{n-1}, \bigwedge_{i=2}^{n-1} AP_1 \equiv AP_i \wedge BP_1 \equiv BP_i \wedge CP_1 \equiv CP_i \wedge$$
$$\neg\beta\,A\,B\,C \wedge \neg\beta\,B\,C\,A \wedge \neg\beta\,C\,A\,B$$

9(1) Dimension, upper bound 1

$$\beta\,A\,B\,C \vee \beta\,B\,C\,A \vee \beta\,C\,A\,B$$

Three points are always on the same line.

9(n) Dimension, upper bound n

$$\bigwedge_{1 \le i < j \le n} P_i \neq P_j \wedge \atop \bigwedge_{i=2}^{n} AP_1 \equiv AP_i \wedge BP_1 \equiv BP_i \wedge CP_1 \equiv CP_i \Rightarrow \beta\,A\,B\,C \vee \beta\,B\,C\,A \vee \beta\,C\,A\,B$$

$9_1(2)$ Dimension, upper bound 2 (variant)[3]

$$\exists Y, (ColXYA \wedge \beta\,B\,Y\,C) \vee (ColXYB \wedge \beta\,C\,Y\,A) \vee (ColXYC \wedge \beta\,A\,Y\,B)$$

10 Euclid's axiom

$$\beta\,A\,D\,T \wedge \beta\,B\,D\,C \wedge A \neq D \Rightarrow \exists X, Y \; \beta\,A\,B\,X \wedge \beta\,A\,C\,Y \wedge \beta\,X\,T\,Y$$

10_1 Euclid's axiom (variant)

$$\beta\,A\,D\,T \wedge \beta\,B\,D\,C \wedge A \neq D \Rightarrow \exists X, Y \; \beta\,A\,B\,X \wedge \beta\,A\,C\,Y \wedge \beta\,Y\,T\,X$$

11 Continuity

$$\exists a, \forall xy, (x \in X \wedge y \in Y \Rightarrow \beta\,a\,x\,y) \Rightarrow \exists b, \forall xy, x \in X \wedge y \in Y \Rightarrow \beta\,x\,b\,y$$

Schema 11 Elementary Continuity (schema)

$$\exists a, \forall xy, (\alpha \wedge \beta \Rightarrow \beta\,a\,x\,y) \Rightarrow \exists b, \forall xy, \alpha \wedge \beta \Rightarrow \beta\,x\,b\,y$$

where α and β are first order formulas, such that a, b and y do not appear free in α; a, b and x do not appear free in β.

---

[3] $ColABC$ is a shorthand for $\beta\,A\,B\,C \vee \beta\,B\,C\,A \vee \beta\,C\,A\,B$ to simplify the presentation. The Col predicate does not belong to the language of the theory of Tarski.

A geometry defined by the elementary continuity axiom schema instead of the higher order continuity axiom is called elementary.

12 Reflexivity of β

$$\beta\,ABB$$

B is always between A and B.

14 Symmetry of β

$$\beta\,ABC \Rightarrow \beta\,CBA$$

If B is between A and C then B is between C and A.

13 Compatibility of equality with β

$$A = B \Rightarrow \beta\,ABA$$

19 Compatibility of equality with $\equiv$

$$A = B \Rightarrow AC \equiv BC$$

15 Transitivity (inner) of β

$$\beta\,ABD \wedge \beta\,BCD \Rightarrow \beta\,ABC$$

16 Transitivity (outer) of β

$$\beta\,ABC \wedge \beta\,BCD \wedge B \neq C \Rightarrow \beta\,ABD$$

17 Connectivity (inner) of β

$$\beta\,ABD \wedge \beta\,ACD \Rightarrow \beta\,ABC \vee \beta\,ACB$$

18 Connectivity (outer) of β

$$\beta\,ABC \wedge \beta\,ABD \wedge A \neq B \Rightarrow \beta\,ACD \vee \beta\,ADC$$

20 Triangle construction unicity

$$
\begin{aligned}
&AC \equiv AC' \wedge BC \equiv BC' \wedge \\
&\beta\,ADB \wedge \beta\,AD'B \wedge \beta\,CDX \wedge \\
&\beta\,C'D'X \wedge D \neq X \wedge D' \neq X
\end{aligned} \Rightarrow C = C'
$$

20_1 Triangle construction unicity (variant)

$$
\begin{aligned}
&A \neq B \wedge \\
&AC \equiv AC' \wedge BC \equiv BC' \wedge \\
&\beta\,BDC' \wedge (\beta\,ADC \vee \beta\,ACD)
\end{aligned} \Rightarrow C = C'
$$

21 Triangle construction existence

$$
AB \equiv A'B' \Rightarrow \exists CX, \quad
\begin{aligned}
&AC \equiv A'C' \wedge BC \equiv B'C' \wedge \\
&\beta\,CXP \wedge (\beta\,ABX \vee \beta\,BXA \vee \beta\,XAB)
\end{aligned}
$$

Year :	1940	1951	1959	1965	1983
Reference :	[18]	[17]	[19]	[20]	[6]
Axioms :	1	1	1	1	1
	2	2	2	2	2
	3	3	3	3	3
	4	4	4	4	4
	5_1	5_1	5	5	5
	6	6	6		6
	7_2	7_2	7_1	7_1	7
	8(2)	8(2)	8(2)	8(2)	8(2)
	$9_1(2)$	$9_1(2)$	9(2)	9(2)	9(2)
	10	10	10_1	10_1	10
	11	11	11	11	11
	12	12			
	13				
	14	14			
	15	15	15	15	
	16	16			
	17	17			
	18	18	18		
	19				
	20	$\rightarrow 20_1$			
	21	21			
Nb of axioms :	20	18	12	10	10
	+	+	+	+	+
	1 schema	1 schema	1 schema	1 schema	1 schema

Fig. 2. History of Tarski's axiom systems

Identity $\beta \, A \, B \, A \Rightarrow (A = B)$

Pseudo-Transitivity $AB \equiv CD \wedge AB \equiv EF \Rightarrow CD \equiv EF$

Symmetry $AB \equiv BA$

Identity $AB \equiv CC \Rightarrow A = B$

Pasch $\exists X, \beta \, A \, P \, C \wedge \beta \, B \, Q \, C \Rightarrow \beta \, P \, x \, B \wedge \beta \, Q \, x \, A$

Euclid $\exists XY, \beta \, A \, D \, T \wedge \beta \, B \, D \, C \wedge A \neq D \Rightarrow$
$\beta \, P \, x \, B \wedge \beta \, Q \, x \, A$

$AB \equiv A'B' \wedge BC \equiv B'C' \wedge$

5 segments $AD \equiv A'D' \wedge BD \equiv B'D' \wedge$
$\beta \, A \, B \, C \wedge \beta \, A' \, B' \, C' \wedge A \neq B \Rightarrow CD \equiv C'D'$

Construction $\exists E, \beta \, A \, B \, E \wedge BE \equiv CD$

Lower Dimension $\exists ABC, \neg\beta \, A \, B \, C \wedge \neg\beta \, B \, C \, A \wedge \neg\beta \, C \, A \, B$

Upper Dimension $AP \equiv AQ \wedge BP \equiv BQ \wedge CP \equiv CQ \wedge P \neq Q$
$\Rightarrow \beta \, A \, B \, C \vee \beta \, B \, C \, A \vee \beta \, C \, A \, B$

Continuity $\forall XY, (\exists A, (\forall xy, x \in X \wedge y \in Y \Rightarrow \beta \, A \, x \, y)) \Rightarrow$
$\exists B, (\forall xy, x \in X \Rightarrow y \in Y \Rightarrow \beta \, x \, B \, y).$

Fig. 3. Tarski's axiom system (Formalized version - 11 axioms)

3.2 History

Tarski began to work on his axiom system in 1926 and presented it during his lectures at Warsaw university[4]. He submitted it for publication in 1940 and was published in his first form in 1967 [18]. This version contains 20 axioms and one schema. A second version, slightly simpler was published in [17]. This first simplification consists only in considering a logic with built-in equality, axioms 13 and 19 are then useless. This second version was further simplified by Eva Kallin, Scott Taylor and Tarski into a system of twelve axioms [19]. The last simplification was obtained by Gupta in its PhD thesis [20], where he gives the proof that two more axioms can be derived from the remaining ones.

Figure 2 gives the list of axioms contained in each of these axiom systems. Figure 3 provides the final list of axioms that we used in our formalization.

4   A Short Introduction to the Coq Proof Assistant

The Coq system [1,21,22] is a proving tool based on a logical formalism called the calculus of inductive constructions [23]. Even if the Coq system has some automatic theorem proving features, it is *not* an automatic theorem prover. The proofs are mainly built by the user *interactively*. The system checks whether the proof is correct. In [14], we have described the formalization of decision procedure for geometry. This formalization, allows to use the area method to generate automatically proofs which are double checked by the Coq system. In this development, we do not want to make use of this procedure. Otherwise we would have a circularity problem because our goal is to provide solid foundations for different formal developments about geometry including this one.

The underlying logic of the Coq system is an intuitionist logic. This means that the proposition $A \vee \neg A$ is not taken for granted and, if it is needed, the user has to assume it explicitly. This allows to clarify the distinction between classical and constructive proofs.

The user interacts with the system using commands which modify the current state of the proof. The language used to interact with the system is called a tactic language[5].

5   Formalization in Coq

The mechanization of the proof we have realized prove formally that the simplifications of the first version of Tarski's axiom system are correct. The unnecessary axioms are derived from the remaining ones.

Now, we provide a quick overview of the content of each chapter. We will only detail an example proof in the next section.

---

[4] We use [5] and the footnotes in [17] to give a quick history of the different versions of Tarski's axiom system.

[5] Note that in the latest version of Coq (8.1) another proof language is available, this new language allows to write proofs which are more readable, unfortunately it was not available when have started this work.

The first chapter contains the declaration of all the axioms and the definition of the collinearity predicate (noted `Col`).

The second chapter contains some basic properties of the equidistance predicate (noted `Cong`). It contains also the proof of the unicity of the point constructed thanks to the segment construction axiom.

The third chapter contains some properties of the betweenness predicate (noted `Bet`). It contains in particular the proof of the axioms 12, 14 and 16.

The fourth chapter contains the proof of several properties of `Cong`, `Col` and `Bet`.

The fifth chapter contains some pseudo-transitivity properties of betweenness and the definition of the length comparison predicate (noted `le`) with some associated properties. It includes in particular the proofs of the axioms 17 and 18.

The sixth chapter defines the `out` predicate which means that a point lies on a line out of a segment. This predicate is used to prove some other properties of `Cong`, `Col` and `Bet` such as transitivity properties for `Col`.

The seventh chapter defines the midpoint of a segment and the symmetric points. It has to be noted that at this step the existence of the midpoint is not derived yet.

The eighth chapter contains the definition of the predicate 'perpendicular' (`Perp`), and the proof of some related properties such as the existence of the foot of the perpendicular. Finally, the existence of the midpoint of a segment is derived.

5.1 Two Crucial Lemmas

Our formalization follows strictly the lines of the book by Schwabhäuser, Szmielew and Tarski except in the fifth chapter where we introduce two crucial lemmas which do not appear in the original text, and which are necessary to fill some gaps in the informal proofs. These two lemmas allows to deduce the equality of two points which lie on a segment under an hypotheses involving distances.

$$\forall ABC, \beta\, A\, B\, C \wedge AC \equiv AB \Rightarrow C = B$$

Proof. We use the lemma 4.6 of [6]:

$$\forall ABCA'B'C', \ \beta\, A\, B\, C \wedge Cong_3 ABCA'B'C' \Rightarrow \beta\, A'\, B'\, C'.$$

As we know by assumption that $\beta\, A\, B\, C$, we apply the lemma with $A := A$, $B := B$, $C := C$, $A' := A$, $B' := C$ and $C' := B$, to obtain that:

$$Cong_3 ABCACB \Rightarrow \beta\, A\, C\, B$$

The predicate $Cong_3 A_1 A_2 A_3 B_1 B_2 B_3$ expresses that:

$$A_1 A_2 \equiv B_1 B_2 \wedge A_1 A_3 \equiv B_1 B_3 \wedge A_2 A_3 \equiv B_2 B_3$$

So here, we need to show that:

$$AB \equiv AC \wedge AC \equiv AB \wedge BC \equiv CB.$$

The first conjunct is shown by commutativity of $\equiv$, the second one by hypothesis and the third one using the pseudo-commutativity property of the oriented distance.

As $\beta\ ABC$ and $\beta\ ACB$, we can conclude that $C = B$ using the lemma between_equality:

$$\forall ABC : Point, \beta\ ABC \wedge \beta\ BAC \Rightarrow A = B$$

and the symmetry property of β .

The second lemma is the following, we omit the proof.

$$\forall ABDE, \beta\ ADB \wedge \beta\ AEB \wedge AD \equiv AE \Rightarrow D = E.$$

5.2   A Comparison Between the Formal and Informal Proofs

We first describe in detail the formal proof of a simple example: the first crucial lemma. Then, we reproduce here one of the non trivial proofs: the proof due to Gupta [20] that axiom 18 can be derived from the remaining ones. We translate the proof from [6] and provide in parallel the mechanized proof as a Coq script. For the conciseness of the presentation we provide only the beginning of the formal proof[6]. For the reader not familiar with the Coq proof assistant, we provide a quick informal explanation of the role of the main tactics we use in these proofs.

intro is used to introduced hypothesis in the context. It is the equivalent of the informal sentence: "Suppose that we have A"

assert is used to state what we want to prove. When it is followed by "..." this means that this assertion can be proved automatically.

DecompExAnd given an existential hypotheses, introduces the witness of the existential and decompose the knowledge about it.

split splits a conjunction into its components. This used the fact fact that to show $A \wedge B$, one can show first A then B.

apply is used to apply an assumption, a lemma or theorem.

Tarski,sTarski,Between,... are automatic tactics which try to prove the current goal. Informally this can be read as "by simple properties of betweenness" or "by direct application of one of the axioms".

unfold replaces something by its definition.

cases_equality perform a reasoning by cases on the equality of two points.

---

[6] The full proofs are available at the address given at the end of this paper.

A First Example. First we input the statement of the lemma in Coq's syntax. $\beta\,ABC$ is noted `Bet A B C` and $AC \equiv AB$ is noted `Conq A C A B`. The text displayed after `Coq <` represents the input, the other pieces of text represent the output by the system.

```
Coq < Lemma between_cong : forall A B C,
Coq <   Bet A C B -> Cong A C A B -> C=B.
Coq < 1 subgoal
```

```
============================
thesis := forall A B C : Point, Bet A C B -> Cong A C A B -> C = B
```

The Coq system confirms that we have one statement to prove. The statements appearing above the horizontal line are the hypotheses and the one appearing under the horizontal line are the fact to be proved. Here we do not have any hypothesis yet. We can now start the proof.

```
Coq < Proof.
```

The first tactic we apply is the **intro** tactic. This introduces in the context the left hand side of the implications.

```
   Coq < intros.
1 subgoal

A : Point
B : Point
C : Point
H : Bet A C B
H0 : Cong A C A B
============================
thesis := C = B
```

Now, we have five hypotheses. We know that A, B and C are points, that $\beta\,ACB$ and that $AC \equiv AB$. We need to show that $C = B$.

Let's first prove that $\beta\,ABC$.

```
   Coq < assert (Bet A B C).
2 subgoals

A : Point
B : Point
C : Point
H : Bet A C B
H0 : Cong A C A B
============================
thesis := Bet A B C
subgoal 2 is: C = B
```

We have two statements two prove. The first one is $\beta\, A\, B\, C$ and the second one is $C = B$. Now, we apply the lemma 4.6. This changes our first subgoal into two subgoals which are the assumptions of the lemma 4.6.

```
Coq < eapply 14_6.
3 subgoals

A : Point
B : Point
C : Point
H : Bet A C B
H0 : Cong A C A B
==============================
thesis := Bet ?X ?Y ?Z
subgoal 2 is: Cong_3 ?X ?Y ?Z A B C
subgoal 3 is: C = B
```

The first subgoal means that we have to prove β $?X$ $?Y$ $?Z$ for some $?X$, $?Y$ and $?Z$. We choose to apply our assumption H. This instantiate $?X$, $?Y$ and $?Z$ with A, C and B respectively.

```
Coq < apply H.
2 subgoals

A : Point
B : Point
C : Point
H : Bet A C B
H0 : Cong A C A B
==============================
thesis := Cong_3 A C B A B C
subgoal 2 is: C = B
```

Now, we need to show that `Cong_3 A C B A B C`. This can be done by unfolding the definition of `Conq_3`, splitting the conjunction we get and solving the resulting subgoals using an automatic tactic.

```
Coq < unfold Cong_3;repeat split;sTarski.
1 subgoal

A : Point
B : Point
C : Point
H : Bet A C B
H0 : Cong A C A B
H1 : Bet A B C
==============================
thesis := C = B
```

Again, we need to show $C = B$, but this time we have the hypothesis H1.

```
Coq < esTarski.
Proof completed.
```

From H and H1 it is possible to conclude using an automatic tactic. Finally, Coq checks again the proof and add it to its database.

```
Coq < Qed.
between_cong is defined
```

Note that during the proof the system checks that the commands we give are correct but in this last step the proof is checked again by a small part of the Coq system called the kernel. Only the kernel of the system needs to be bug free to ensure the correctness of the proof. Bugs which are outside the kernel can not lead to a proof of a false statement.

Axiom 18

Theorem 1 (Gupta). $A \neq B \wedge \beta\, ABC \wedge \beta\, ABD \Rightarrow \beta\, ACD \vee \beta\, ADC$

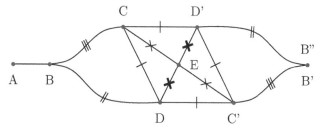

Proof: Let C' and D' be points such that :

$$\beta\, ADC' \wedge DC' \equiv CD \text{ and } \beta\, ACD' \wedge CD' \equiv CD$$

```
assert (exists C', Bet A D C' /\ Cong D C' C D)...
DecompExAnd H2 C'.
assert (exists D', Bet A C D' /\ Cong C D' C D)...
DecompExAnd H2 D'.
```

We have to show that $C = C'$ or $D = D'$.
Let B and B'' points such that :

$$\beta\, AC'B' \wedge C'B' \equiv CB \text{ and } \beta\, AD'B'' \wedge D'B'' \equiv DB$$

```
assert (exists B', Bet A C' B' /\ Cong C' B' C B)...
DecompExAnd H2 B'.
assert (exists B'', Bet A D' B'' /\ Cong D' B'' D B)...
DecompExAnd H2 B''.
```

Using the lemma 2.11[7] we can deduce that $BC' \equiv B''C$ and that $BB' \equiv B''B$.

```
assert (Cong B C' B'' C).
eapply 12_11.
3:apply cong_commutativity.
3:apply cong_symmetry.
3:apply H11.
Between.
Between.
esTarski.
assert (Cong B B' B'' B).
eapply 12_11;try apply H2;Between.
```

By unicity of the segment construction, we know that $B'' = B'$.

```
assert (B''=B').
apply construction_unicity with
(Q:=A) (A:=B) (B:=B'') (C:=B) (x:=B'') (y:=B');Between...
smart_subst B''.
```

We know that $FSC \begin{pmatrix} BCD'C' \\ B'C'DC \end{pmatrix}$ (The points form a five segments configuration).

```
assert (FSC B C D' C' B' C' D C).
unfold FSC;repeat split;unfold Col;Between;sTarski.
2:eapply cong_transitivity.
2:apply H7.
2:sTarski.
apply 12_11 with (A:=B) (B:=C) (C:=D') (A':=B') (B':=C') (C':=D);
Between;sTarski;esTarski.
```

Hence $C'D' \equiv CD$ (because if $B \neq C$ the five segments axiom gives the conclusion and if $B = C$ we can use the hypotheses).

```
assert (Cong C' D' C D).
cases_equality B C.
(* First case *)
treat_equalities.
eapply cong_transitivity.
apply cong_commutativity.
apply H11.
Tarski.
(* Second case *)
apply cong_commutativity.
eapply 14_16;try apply H3...
```

---

[7] The lemma 2.11 states that $\beta\,ABC \wedge \beta\,A'B'C' \wedge AB \equiv A'B' \wedge BC \equiv B'C' \Rightarrow AC \equiv A'C'$.

Using the axiom of Pasch, there is a point E such that :

$$\beta\, C\, E\, C' \wedge \beta\, D\, E\, D'$$

```
assert (exists E, Bet C E C' /\ Bet D E D').
eapply inner_pash;Between.
DecompExAnd H13 E.
```

We omit the rest of the formal proof.

We can deduce that $IFS\left(\dfrac{ded'c}{ded'c'}\right)$ and $IFS\left(\dfrac{cec'd}{cec'd'}\right)$. Hence $EC \equiv EC'$ and

$ED \equiv ED'$. Suppose that $C \neq C'$. We have to show that $D = D'^8$. From the hypotheses, we can infer that $C \neq D'$. Using the segment construction axiom, we know that there are points P, Q and R such that :

$$\beta\, C'\, C\, P \wedge CP \equiv CD' \text{ and } \beta\, D'\, C\, R \wedge CR \equiv CE \text{ and } \beta\, P\, R\, Q \wedge RQ \equiv RP$$

Hence $FSC\left(\dfrac{D'CRP}{PCED'}\right)$, so $RP \equiv ED'$ and $RQ \equiv ED$. We can infer that

$FSC\left(\dfrac{D'EDC}{PRQC}\right)$, so using lemma 2.11 we can conclude that $D'D \equiv PQ$ and $CQ \equiv CD$ (because the case $D' \neq E$ is solved using the five segments axiom, and in the other case we can deduce that $D' = D$ and $P = Q$). Using the theorem 4.17[9], as $R \neq C$ and R, C and D' are collinear we can conclude that $D'P \equiv D'Q$. As $C \neq D'$, $Col\,CD'B$ and $Col\,CD'B'$, we can also deduce that $BP \equiv BQ$ and $B'P \equiv B'Q$. As $C \neq D'$, we have $B \neq B'$ and as $Col\,BC'B'$ we have $C'P \equiv C'Q$. As $C \neq C'$ and $Col\,C'CP$ we have $PP \equiv PQ$. Using the identity axiom for equidistance, we can deduce that $P = Q$. As $PQ \equiv D'D$, we also have $D = D'$. □

5.3 About Degenerated Cases

Every paper about the formalization of geometry, in particular those about Hilbert's foundations of geometry [2,4] emphasizes the problem of the degenerated cases. In geometry, the degenerated cases are limit cases such as when two points are equals, three points are collinear or two lines are parallel. The formal proof of the theorems in the degenerated cases is often tedious and even sometimes difficult. These cases often do not even appear in the informal proof[10]. In order to limit the size of the proofs, we tried to automate some tasks. These pieces of automation should not be compared with the highly successful decision procedures for geometry, the goal is just to automate some easy but very tedious proofs

---

[8] Note that this step uses the decidability of equality between two points.

[9] The theorem 4.17 states that $A \neq B \wedge Col\,ABC \wedge AP \equiv AQ \wedge BP \equiv BQ \Rightarrow CP \equiv CQ$.

[10] It seems that degenerated cases play the same role in geometry as α-conversion in lambda calculus: they are a great source of difficulties in the context of a mechanization.

and, as stated before, as our goal is to build foundations for the implementation of decision procedures we can not use these more powerful procedures.

The main tactic to deal with degenerated cases is called `treat_equalities`. The basic idea is to propagate information about degenerated cases. For instance, if we know that $A = B$ and $AB \equiv CD$ we can deduce that $C = D$. This is very simple but it shortens the proofs of the degenerated cases quite effectively.

Moreover, we think that a source of degenerated cases come from the axiom system. In our personal experience the formalization of geometry using Hilbert axioms lead to far more degenerated cases because the axioms are not always stated in the most general and uniform way. We think that Tarski's geometry is a good candidate to mechanization because it is very simple, it has good meta-mathematical properties (cf [17]) and it produces few degenerated cases.

5.4   Comparison with Other Formalizations

Compared to Frédérique Guilhot formalization [10], our development should be considered low level. Our formalization has the advantage of being based on the axiom system of Tarski which is of an extreme simplicity: two predicates and eleven axioms. But this simplicity has a price, our formalization is not adapted to the context of education. Indeed, some intuitively simple properties are hard to prove in this context. For instance, the proof of the existence of the midpoint of segment is obtained only at the end of the eighth chapter after about 150 lemmas and 4000 lines of proof. Moreover, the small number of axioms imposes a scheduling of the lemmas which is not always intuitive. Indeed, some simple intuitive properties can only be proved late in the development. For instance the transitivity properties for collinearity are only proved in the chapter 6, this means that in the first fifth chapters we have to live in a world where we do not assume that collinearity has some transitivity properties.

Compared with formalizations using Hilbert's axiom system, we think that, as stated in the previous section, the use of Tarski's axiom system leads to more uniform proofs with less degenerated cases. Note that there are degenerated cases which are inherent to a statement: the statement is false otherwise. There are also degenerated cases which are inherent to the formulation of a statement, if one starts with an axiom system which contains numerous degenerated cases then the proofs of the first lemmas have to deal with these cases to obtain more uniform statements. The use of Tarski's axiom system has also the advantage that, as it is based only on points, it can be easily generalized to other dimensions by just changing the dimension axiom. In practice, in the context of a formal proof, this allows to prove the lemmas which do not use the dimension axiom only once. On the other end using Hilbert's axiom system, to change the dimension of the space, the language and axioms have to be changed and the proofs as well, for dimension 3 for instance, it is necessary to assume the existence of planes.

5.5   Classical vs. Intuitionist Logic

Our formalization of Tarski's geometry is performed in the system Coq. As the logic behind Coq is constructive, we need to tell Coq explicitly when we need classical logic. This is the case in this development. It appears quite often in the proofs that we need to distinguish between two cases such that $A = B$ and $A \neq B$ or $Col\,ABC$ and $\neg Col\,ABC$. This kind of reasoning relies on the decidability of point equality and collinearity. We proved these two facts using the excluded middle rule.

6   Future Work and Conclusion

A natural extension of our work consist in mechanizing the remaining chapters of [6] and proving the axioms of Hilbert. This work is under progress. We also plan to enrich our formalization to use it as a foundation for other formal Coq developments about geometry such as Frédérique Guilhot formalization of geometry as it is presented in the french curriculum [10] and our implementation in Coq of the area method of Chou, Gao and Zhang [14]. A longer-term challenge would be to perform a systematic development of geometry similar to the book of Schwabhäuser, Szmielew and Tarski but in the context of a constructive axiom system such as the axiom system of von Plato [24] which has already been formalized in the Coq proof assistant by Gilles Khan [25].

We have presented the mechanization of the proofs of over 150 lemmas in the context of Tarski's geometry. This includes the formal proof that the simplifications of the first version of Tarski's axiom system are corrects. Our main conclusion is that Tarski axiom system lead to more uniform proofs than Hilbert's axiom system and so it is better suited for a formalization.

Availability

The full Coq development with the formal *proofs* and hypertext links to ease navigation can be found at the following url:

http://www.lix.polytechnique.fr/Labo/Julien.Narboux/tarski.html

References

1. Coq development team, The: The Coq proof assistant reference manual, Version 8.0. LogiCal Project (2004)
2. Dehlinger, C., Dufourd, J.F., Schreck, P.: Higher-order intuitionistic formalization and proofs in Hilbert's elementary geometry. In: Richter-Gebert, J., Wang, D. (eds.) ADG 2000. LNCS (LNAI), vol. 2061, pp. 306–324. Springer, Heidelberg (2001), http://www.springerlink.com/content/p2mh1ad6ede09g6x/
3. Paulson, L.C.: The Isabelle reference manual (2006)
4. Meikle, L., Fleuriot, J.: Formalizing Hilbert's Grundlagen in Isabelle/Isar. In: Basin, D., Wolff, B. (eds.) TPHOLs 2003. LNCS, vol. 2758, pp. 319–334. Springer, Heidelberg (2003),
http://springerlink.metapress.com/content/6ngakhj9k7qj71th/?p=e4d2c272843b41148f0550f9c5ad8a2a&pi=20
5. Tarski, A., Givant, S.: Tarski's system of geometry. The bulletin of Symbolic Logic 5(2) (1999)

6. Schwabhäuser, W., Szmielew, W., Tarski, A.: Metamathematische Methoden in der Geometrie. Springer, Berlin (1983)
7. Quaife, A.: Automated development of Tarski's geometry. Journal of Automated Reasoning 5(1), 97–118 (1989)
8. Narboux, J.: Toward the use of a proof assistant to teach mathematics. In (ICTMT7). Proceedings of the 7th International Conference on Technology in Mathematics Teaching (2005)
9. Narboux, J.: A graphical user interface for formal proofs in geometry. The Journal of Automated Reasoning special issue on User Interface for Theorem Proving 39(2) (2007), http://www.informatik.uni-bremen.de%7Ecxl/uitp-jar/, http://springerlink.metapress.com/content/b60418176x313vx6/? p=aa2ae37923ab455291ee15f6c9c39b9b&pi=3
10. Guilhot, F.: Formalisation en coq et visualisation d'un cours de géométrie pour le lycée. Revue des Sciences et Technologies de l'Information, Technique et Science Informatiques, Langages applicatifs 24, 1113–1138 (2005)
11. Pichardie, D., Bertot, Y.: Formalizing convex hulls algorithms. In: Boulton, R.J., Jackson, P.B. (eds.) TPHOLs 2001. LNCS, vol. 2152, pp. 346–361. Springer, Heidelberg (2001)
12. Meikle, L., Fleuriot, J.: Mechanical theorem proving in computation geometry. In: Hong, H., Wang, D. (eds.) ADG 2004. LNCS (LNAI), vol. 3763, pp. 1–18. Springer, Heidelberg (2006)
13. Dufourd, J.F.: Design and formal proof of a new optimal image segmentation program with hypermaps. In: Pattern Recognition, vol. 40(11), Elsevier, Amsterdam (2007), http://portal.acm.org/citation.cfm?id=1274191.1274325&coll= GUIDE&dl=%23url.coll
14. Narboux, J.: A decision procedure for geometry in Coq. In: Slind, K., Bunker, A., Gopalakrishnan, G.C. (eds.) TPHOLs 2004. LNCS, vol. 3223, Springer, Heidelberg (2004)
15. Narboux, J.: Formalisation et automatisation du raisonnement géométrique en Coq. PhD thesis, Université Paris Sud (September 2006)
16. Chou, S.C., Gao, X.S., Zhang, J.Z.: Machine Proofs in Geometry. World Scientific, Singapore (1994)
17. Tarski, A.: A decision method for elementary algebra and geometry. University of California Press (1951)
18. Tarski, A.: The completeness of elementary algebra and geometry (1967)
19. Tarski, A.: What is elementary geometry? In: Henkin, L., Tarski, P.S. (eds.) The axiomatic Method, with special reference to Geometry and Physics, Amsterdam, North-Holland, pp. 16–29 (1959)
20. Gupta, H.N.: Contributions to the axiomatic foundations of geometry. PhD thesis, University of California, Berkley (1965)
21. Huet, G., Kahn, G., Paulin-Mohring, C.: The Coq Proof Assistant - A tutorial - Version 8.0 (April 2004)
22. Bertot, Y., Castéran, P.: Interactive Theorem Proving and Program Development, Coq'Art: The Calculus of Inductive Constructions. In: Texts in Theoretical Computer Science. An EATCS Series, Springer, Heidelberg (2004)
23. Coquand, T., Paulin-Mohring, C.: Inductively defined types. In: Martin-Löf, P., Mints, G. (eds.) COLOG-88. LNCS, vol. 417, Springer, Heidelberg (1990)
24. von Plato, J.: The axioms of constructive geometry. Annals of Pure and Applied Logic 76, 169–200 (1995)
25. Kahn, G.: Constructive geometry according to Jan von Plato. Coq contribution, Coq V5.10 (1995)

On the Need of Radical Ideals in Automatic Proving: A Theorem About Regular Polygons

Pavel Pech

Pedagogical Faculty, University of South Bohemia,
371 15 České Budějovice, Czech Republic
pech@pf.jcu.cz

Abstract. The paper deals with a problem of finding natural geometry problem, that is, not specifically built up for the only purpose of having some concrete property, where the hypothesis is not described by a radical ideal. This problem was posed by Chou long ago. Regular polygons in the Euclidean space E^d and their existence in spaces of various dimensions are studied by the technique of Gröbner bases. When proving that regular pentagons and heptagons span spaces of even dimension one encounters the case that the ideal describing the hypotheses is not radical. Thus, in order to prove that $H \Rightarrow T$ one needs to show that T belongs to the radical of the ideal describing H.

1 Introduction

In the paper regular polygons in the Euclidean space E^d are studied by the method which is based on Gröbner bases computation. Properties of regular pentagons and regular heptagons, especially their existence in spaces of various dimensions, are investigated (Theorem 1, Theorem 2). In addition, a sufficient and necessary condition for a heptagon in E^d to be planar is given (Theorem 3). The novelty of the paper are the proofs of Theorems 1, 2 filling a gap as to the problem posed by Chou, the Theorem 3 seems to be new.

Given a set H of hypotheses $h_1 = 0, h_2 = 0, \ldots, h_r = 0$ we are to prove that from these hypotheses the conclusion T, given by the equation $c = 0$, follows. By the theory of automatic theorem proving, in accordance with the Hilbert's Nullstellensatz [9], we are to show that c belongs to the radical $\sqrt{I}$, where we denote $I = (h_1, h_2, \ldots, h_r)$. In practice we often encounter the case $\sqrt{I} = I$ which allows to show that c is an element of the ideal I. S. Ch. Chou in his well-known book Mechanical Geometry Theorem Proving [8], where over five hundred examples are given, on the page 78 writes that " ... for all theorems we have found in practice, $I = \sqrt{I}$ ". Exploring properties of regular polygons in E^d we can meet the case in which it is necessary to show that the conclusion polynomial belongs to the radical ideal $\sqrt{I}$. Regular polygons yield a suitable topic to investigate them from this point of view. Let us give a brief introduction into the interesting history of regular polygons [21].

Before Christmas 1969 two chemists A. Dreiding and J. D. Dunitz visited the well-known mathematician B. L. van der Waerden. The latter talked about fixed

F. Botana and T. Recio (Eds.): ADG 2006, LNAI 4869, pp. 157–170, 2007.

and movable forms of cyclic hexane and octane. He also mentioned cyclic pentane and insisted that a skew pentagon with the same side lengths and the same angles must necessary lie in a plane. Van der Waerden was surprised at this statement and asked for explanation. Already on the 10th February 1970 at the Mathematics Colloquium in Zürich B.L. van der Waerden held a lecture "Ein Satz über räumliche Fünfecke" whose only topic was a proof of the following theorem [21]:

A skew pentagon with equal sides and equal angles is planar
After the publication of the paper [21] this theorem was proved in a short time by a number of mathematicians in a few ways, see [6], [15], [22]. B. Grünbaum [11] writes that this property of a regular pentagon — i.e. equilateral and equiangular — was known to A. Auric [1] in 1911. Detailed information is described in the survey article [11], where it is stated that a total characterization of regular n-gons was done in 1962 by Russians V. A. Efremovitch and Ju. S. Iljjashenko [10].

2   Preliminaries

In this section we give some notions both from the theory of polygons and from the theory of automatic theorem proving that we will need in the next part. Remember that all subscripts are taken modulo n.

A polygon $P_0, P_1, \ldots, P_{n-1}$ whose sides have the same length, i.e. $|P_j P_{j+1}|$ is constant for all $j = 0, 1, \ldots, n-1$, we call *equilateral*. Similarly, an n-gon is *k-equilateral* if $|P_j P_{j+\nu}| = d_\nu$ for all $j = 0, 1, \ldots, n-1$ and $\nu = 1, 2, \ldots, k$, where the constants d_ν are *parameters*.
The definition of a regular polygon is as follows [11]:

A polygon $P_0, P_1, \ldots, P_{n-1}$ is called regular if for all $\nu = 1, 2, \ldots, n-1$ the lengths of segments $P_j P_{j+\nu}$ are independent of j, or in other words, if a polygon is $(n-1)$-equilateral.

Thus an equilateral n-gon is 1-equilateral with the parameter d_1, 2-equilateral n-gon is equilateral and equiangular with parameters d_1, d_2, etc. If we introduce $d_0 = 0$ then a regular n-gon is characterized by relations

$$|P_j P_{j+\nu}| = d_\nu, \quad \text{for all} \quad j, \nu = 0, 1, \ldots, n-1,$$

which means that all diagonals of the "same" kind (next but one vertex, next but two vertices, ...) have the same length.

The volume V_n of a simplex $A_1 A_2 \ldots A_{n+1}$ in E^n can be expressed in terms of all mutual distances $|A_i A_j| = a_{ij}$ between vertices of a simplex in the form of the so called Cayley–Menger's determinant [2]:

Denoting $a_{ij}^2 = d_{ij}$ then for the volume V_n of a simplex $A_1, A_2, \ldots, A_{n+1}$

$$(-1)^{n+1} 2^n (n!)^2 V_n^2 = D_n = \begin{vmatrix} 0 & 1 & 1 & 1 & \ldots & 1 \\ 1 & 0 & d_{12} & d_{13} & \ldots & d_{1,n+1} \\ 1 & d_{21} & 0 & d_{23} & \ldots & d_{2,n+1} \\ \ldots & \ldots & \ldots & \ldots & \ldots & \ldots \\ 1 & d_{n+1,1} & d_{n+1,2} & \ldots & \ldots & 0 \end{vmatrix}. \tag{1}$$

We say that an n-gon has *dimension s*, if s is dimension of the least subspace of an affine space A^d in which an n-gon is involved or, which is equivalent, if an n-gon spans a s-dimensional space.

We shall investigate the dimension of a regular polygon in E^d by means of Gröbner bases computation in automated theorem proving. Let us give basic definitions and theorems [9], [12], [18], [19], [23].

Let K be a field of characteristic 0, for instance the field of rational numbers $\mathbb{Q}$, and L an algebraically closed field containing K, for instance the field of complex numbers $\mathbb{C}$. Further denote by $K[x_1, \ldots, x_n]$ the ring of polynomials of n indeterminates $x = (x_1, \ldots, x_n)$ with coefficients in the field K.

Automatic theorem proving deals with geometric statements which are of the form $H \Rightarrow T$, where H is a set hypotheses

$$h_1(x) = 0, h_2(x) = 0, \ldots, h_r(x) = 0$$

and T is a set of conclusions (theses)

$$c_1(x) = 0, c_2(x) = 0, \ldots, c_s(x) = 0,$$

where $h_1, \ldots, h_r, c_1, \ldots, c_s \in K[x_1, \ldots, x_n]$. Without loss of generality assume that the set of conclusions T contains only one conclusion which we denote by c, i.e.

$$c(x) = 0.$$

The algebraic form of a statement $H \Rightarrow T$ is as follows

$$\forall x \in L^n, \quad h_1(x) = 0, \ldots, h_r(x) = 0 \quad \Rightarrow \quad c(x) = 0. \tag{2}$$

We say that a statement (2) is *generally true* if the hypotheses variety $V(H) = \{x; \ h_1(x) = 0, \ldots, h_r(x) = 0\}$ is a subset of the conclusion variety $V(T) = \{x; \ c(x) = 0\}$.

The next theorem gives an instruction how to show that (2) is generally true [18]:

The following statements are equivalent:

a) The statement (2) is generally true,

b) $c \in \sqrt{(h_1, \ldots, h_r)}$,

c) $1 \in (h_1, \ldots, h_r, ct - 1)$.

Thus to prove that (2) is generally true it suffices to show that the constant polynomial 1 lies in the ideal

$$J = (h_1, \ldots, h_r, ct - 1),$$

where t is a slack variable. By the previous theorem we determine the normal form of 1 with respect to the Gröbner basis of the ideal $J = (h_1, \ldots, h_r, ct - 1)$ for some prescribed order of variables. In the computer algebra system CoCoA[1]

---

[1] Software CoCoA is freely distributed at the address http://cocoa.dima.unige.it

we use the command NF(1,J). If the answer is 0 then it means that $1 \in J$ and (2) is generally true.

Another way how to find out whether the constant 1 belongs to the ideal J is to compute directly the Gröbner basis of the ideal J using any term order. This is useful especially in the case if we use the software which does not automatically compute the normal form using the Gröbner basis of a set of generators.

If we do not manage to prove that the statement (2) is generally true, it still does not mean that the statement is not valid. It can happen that a statement is not generally true because of missing *non-degeneracy conditions*

$$g_1(x_1, \ldots, x_n) \neq 0, \ldots, \ g_s(x_1, \ldots, x_n) \neq 0. \tag{3}$$

We say that the statement (2) is *generically true* if the hypotheses variety $V(H') = \{x; \ h_1(x) = 0, \ldots, h_r(x) = 0, g_1(x) \neq 0, \ldots, g_s(x) \neq 0\}$ is a subset of the conclusion variety $V(T) = \{x; \ c(x) = 0\}$.

In order to prove a statement

$$\forall x \in L^n, h_1(x) = 0, \ldots, h_r(x) = 0, g_1(x) \neq 0, \ldots, g_s(x) \neq 0 \Rightarrow c(x) = 0. \tag{4}$$

it suffices to show, in accordance with the above theorem, that the constant polynomial 1 belongs to the ideal

$$I' = (h_1, \ldots, h_r, g_1 t_1 - 1, \ldots, g_s t_s - 1, ct - 1),$$

where $t_1, \ldots, t_s, t$ are slack variables.

Most of valid statements are generically true. Non-degeneracy conditions of investigated objects are usually not involved in the statements. We tacitly suppose that a triangle is a real triangle and not a segment, that a segment is a real segment and not a pair of identical points, that a circle has the non-zero radius, etc. We say that such objects are *generic*.

Remark 1. In practice it is mostly sufficient to show that the conclusion polynomial c is in the ideal $I = (h_1, \ldots, h_r)$ (Ideal Membership). The normal form of a polynomial c with respect to the Gröbner basis of the ideal I can be found by a command NF(c,I). If we get the answer 0 then a statement is generally true. Otherwise we have to apply the stronger criterion and find out whether 1 is an element of the ideal $J = I \cup \{ct - 1\}$ (Radical Membership).

In the next part we will give an example of $I \subsetneq \sqrt{I}$ when c is not in the ideal I, but it is an element of the radical $\sqrt{I}$.

3   Regular Pentagon

Consider the theorem [21]:

Theorem 1. *A regular skew pentagon $ABCDE$ in the Euclidean space E^3 is given. Then $ABCDE$ is a planar pentagon.*

We will prove this theorem using the theory of automated theorem proving. Assume that a pentagon is equilateral with the side length a and equiangular with the length of a diagonal u.

Let us introduce a Cartesian coordinate system such that for the vertices of a pentagon $A = [a, 0, 0]$, $B = [b_1, b_2, 0]$, $C = [c_1, c_2, c_3]$, $D = [d_1, d_2, d_3]$, $E = [0, 0, 0]$, Fig. 1. Then the following relations are fulfilled:

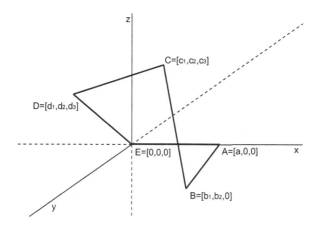

Fig. 1. Regular pentagon in E^3 is planar

$|AB| = a \Leftrightarrow h_1 : (b_1 - a)^2 + b_2^2 - a^2 = 0,$
$|BC| = a \Leftrightarrow h_2 : (c_1 - b_1)^2 + (c_2 - b_2)^2 + c_3^2 - a^2 = 0,$
$|CD| = a \Leftrightarrow h_3 : (d_1 - c_1)^2 + (d_2 - c_2)^2 + (d_3 - c_3)^2 - a^2 = 0,$
$|DE| = a \Leftrightarrow h_4 : d_1^2 + d_2^2 + d_3^2 - a^2 = 0,$
$|AC| = u \Leftrightarrow h_5 : (c_1 - a)^2 + c_2^2 + c_3^2 - u^2 = 0,$
$|BD| = u \Leftrightarrow h_6 : (d_1 - b_1)^2 + (d_2 - b_2)^2 + d_3^2 - u^2 = 0,$
$|CE| = u \Leftrightarrow h_7 : c_1^2 + c_2^2 + c_3^2 - u^2 = 0,$
$|DA| = u \Leftrightarrow h_8 : (d_1 - a)^2 + d_2^2 + d_3^2 - u^2 = 0,$
$|EB| = u \Leftrightarrow h_9 : b_1^2 + b_2^2 - u^2 = 0.$

The points A, B, C, D, E are complanar $\Leftrightarrow A, B, C, E$ and A, B, D, E are complanar $\Leftrightarrow$

$$
z_1 : \begin{vmatrix} 0 & 0 & 0 & 1 \\ a & 0 & 0 & 1 \\ b_1 & b_2 & 0 & 1 \\ c_1 & c_2 & c_3 & 1 \end{vmatrix} = 0 \quad \text{and} \quad z_2 : \begin{vmatrix} 0 & 0 & 0 & 1 \\ a & 0 & 0 & 1 \\ b_1 & b_2 & 0 & 1 \\ d_1 & d_2 & d_3 & 1 \end{vmatrix} = 0. \tag{5}
$$

We shall explore whether both conclusion polynomials z_1, z_2 belong to the radical of $I = (h_1, h_2, \ldots, h_9)$. We will investigate each polynomial z_1, z_2 separately. For z_1 we find out whether $1 \in J$, where $J = I \cup \{ab_2c_3t - 1\}$. We get

```
Use R::=Q[aub[1..3]c[1..3]d[1..3]t];
J:=Ideal((b[1]-a)^2+b[2]^2-a^2,(c[1]-b[1])^2+(c[2]-b[2])^2+c[3]^2
-a^2,(d[1]-c[1])^2+(d[2]-c[2])^2+(d[3]-c[3])^2-a^2,d[1]^2+d[2]^2+
```

```
d[3]^2-a^2,(c[1]-a)^2+c[2]^2+c[3]^2-u^2,(d[1]-b[1])^2+(d[2]-
b[2])^2+d[3]^2-u^2,c[1]^2+c[2]^2+c[3]^2-u^2,(d[1]-a)^2+d[2]^2+
d[3]^2-u^2,b[1]^2+b[2]^2-u^2,ab[2]c[3]t-1);
NF(1,J);
```

that for the normal form NF(1,J)=0 which means that the points A, B, C, E are complanar.

Similarly we will show that the points A, B, D, E are complanar as well. Whence we can conclude — a regular skew pentagon $ABCDE$ is planar.

In the last proof we examined the normal form NF(1,J), where the ideal J contained the negated conclusion polynomial $ab_2c_3t - 1$. The result NF(1,J)=0 means that the conclusion polynomial ab_2c_3 is an element of the radical $\sqrt{I}$ from which $ab_2c_3 = 0$ follows. Usually it suffices to find out whether the conclusion polynomial ab_2c_3 belongs to the ideal I. Let us do it. We get

```
Use R::=Q[aub[1..3]c[1..3]d[1..3]];
I:=Ideal((b[1]-a)^2+b[2]^2-a^2,(c[1]-b[1])^2+(c[2]-b[2])^2+c[3]^2
-a^2,(d[1]-c[1])^2+(d[2]-c[2])^2+(d[3]-c[3])^2-a^2,d[1]^2+d[2]^2+
d[3]^2-a^2,(c[1]-a)^2+c[2]^2+c[3]^2-u^2,(d[1]-b[1])^2+(d[2]-
b[2])^2+d[3]^2-u^2,c[1]^2+c[2]^2+c[3]^2-u^2,(d[1]-a)^2+d[2]^2+
d[3]^2-u^2,b[1]^2+b[2]^2-u^2);
NF(ab[2]c[3],I);
```

the non zero result ab_2c_3. Hence the polynomial ab_2c_3 *does not* belong to the ideal I. However, ab_2c_3 belongs to the radical $\sqrt{I}$, i.e. there exists a natural number m such that $(ab_2c_3)^m$ belongs to the ideal I. It is easy to verify that $m = 3$, i.e. $(ab_2c_3)^3 \in I$.

Another proof of the fact that a regular pentagon in a space E^3 is planar, is based on the well-known theorem about the volume of a simplex expressed by the Cayley–Menger's determinant (1). This kind of proof will be used later in the case of a regular heptagon.

We will prove the Theorem 1 in the generalized form [6]:

A regular pentagon $A_1A_2A_3A_4A_5$ lies either in E^4 or in E^2, i.e. its dimension is 4 or 2.

The proof, which is due to O. Bottema [6], is as follows.

If we put $V_n = 0$ then by (1) also $D_n = 0$ and points $A_1, A_2, \ldots, A_{n+1}$ span a space whose dimension is less than n.

Minimal dimension of a space in which an arbitrary pentagon can be considered is four. Let $A_1A_2A_3A_4A_5$ be a regular pentagon in an Euclidean space E^4. The vertices of a pentagon in E^4 form a simplex whose volume V_4 can be expressed in terms of all distances between the vertices of a simplex. Denote the lengths of its sides and diagonals by

$$|A_1A_2| = |A_2A_3| = |A_3A_4| = |A_4A_5| = |A_5A_1| = a,$$

$$|A_1A_3| = |A_3A_5| = |A_5A_2| = |A_2A_4| = |A_4A_1| = b.$$

By (1) for a simplex $A_1 A_2 A_3 A_4 A_5$ in E^4 it holds

$$
D_4 =
\begin{vmatrix}
0 & 1 & 1 & 1 & 1 & 1 \\
1 & 0 & a^2 & b^2 & b^2 & a^2 \\
1 & a^2 & 0 & a^2 & b^2 & b^2 \\
1 & b^2 & a^2 & 0 & a^2 & b^2 \\
1 & b^2 & b^2 & a^2 & 0 & a^2 \\
1 & a^2 & b^2 & b^2 & a^2 & 0
\end{vmatrix}
= -5(a^2 + ab - b^2)^2(a^2 - ab - b^2)^2.
\tag{6}
$$

Suppose that $D_4 = 0$. This means that dimension of the pentagon $A_1 A_2 A_3 A_4 A_5$ is less than or equal to three.

Now consider four vertices of $A_1 A_2 A_3 A_4 A_5$, for instance A_1, A_2, A_3, A_4, which form a simplex in E^3 and explore its volume. By (1) we omit the last row and column in the determinant (6). We get

$$
D_3 =
\begin{vmatrix}
0 & 1 & 1 & 1 & 1 \\
1 & 0 & a^2 & b^2 & b^2 \\
1 & a^2 & 0 & a^2 & b^2 \\
1 & b^2 & a^2 & 0 & a^2 \\
1 & b^2 & b^2 & a^2 & 0
\end{vmatrix}
= -2(a^2 + ab - b^2)(a^2 - ab - b^2)(a^2 + b^2).
\tag{7}
$$

Hence

$$
D_4 = 0 \quad \Rightarrow \quad D_3 = 0.
\tag{8}
$$

In order to prove (8) algebraically, we necessarily need to show that D_3 belongs to the radical ideal $\sqrt{D_4}$ since the polynomial D_3 *is not* an element of the ideal generated by D_4. We can see this without the use of a computer.

The same result we obtain for remaining quadruples of the vertices of a pentagon. Thus, if dimension of the pentagon $A_1 A_2 A_3 A_4 A_5$ is not 4 then it is neither 3. Therefore the pentagon $A_1 A_2 A_3 A_4 A_5$ must be planar.

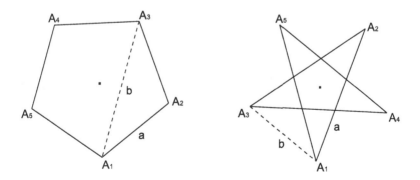

Fig. 2. Convex and non-convex regular pentagons

Remark 2. A regular pentagon $A_1A_2A_3A_4A_5$ in E^2 with the length of side a and diagonal b obeys by (6) and (7) the equation

$$(a^2 + ab - b^2)(a^2 - ab - b^2) = 0. \tag{9}$$

The equation $a^2 + ab - b^2 = 0$ holds for a convex regular pentagon, Fig. 2 on the left, from which $a/b = (-1 + \sqrt{5})/2 = 0.618...$ follows.
From $a^2 - ab - b^2 = 0$ we get $a/b = (1 + \sqrt{5})/2 = 1.618...$ which characterizes non-convex star regular pentagons, Fig. 2 on the right.
From (6) and (7) we get the theorem [6]:

A regular pentagon with the length of sides a and diagonals b is planar if and only if the condition (9) holds. The equation $a^2 + ab - b^2 = 0$ gives a convex regular pentagon, whereas $a^2 - ab - b^2 = 0$ leads to a non-convex regular pentagon.

4   Regular Heptagon

The case of a regular pentagon from the previous part is well-known and has been published many times. On the other hand a regular heptagon and its existence in spaces of various dimensions has not been mentioned explicitly so many times as a regular pentagon. This is likely due to the following general theorem on regular polygons with an arbitrary number of vertices [4], [10], [11]:

A regular polygon with an odd number of vertices has even dimension.

Let us look at a regular heptagon in detail. We will investigate its properties using the method which is based on Gröbner bases computation. We will apply the method which was used by a regular pentagon and which is based on the Cayley–Menger determinant.

Let $A_1A_2A_3A_4A_5A_6A_7$ be a regular heptagon. In accordance with the definition a regular heptagon is 3-equilateral. Here we have to pay attention to the

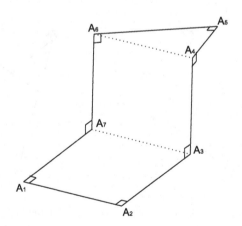

Fig. 3. An equilateral and equiangular heptagon which is not regular

fact that the conditions being equilateral and equiangular, which were sufficient for a pentagon to be regular, do not suffice for a regular heptagon. For instance in Fig. 3 we see an equilateral and equiangular heptagon which is not regular since $|A_7A_3| \neq |A_1A_4|$, see [11].

In a regular heptagon $A_1A_2A_3A_4A_5A_6A_7$ let us denote

$$|A_1A_2| = |A_2A_3| = |A_3A_4| = |A_4A_5| = |A_5A_6| = |A_6A_7| = |A_7A_1| = a,$$

$$|A_1A_3| = |A_3A_5| = |A_5A_7| = |A_7A_2| = |A_2A_4| = |A_4A_6| = |A_6A_1| = b,$$

$$|A_1A_4| = |A_4A_7| = |A_7A_3| = |A_3A_6| = |A_6A_2| = |A_2A_5| = |A_5A_1| = c.$$

Consider a heptagon in a six dimensional Euclidean space E^6 which is a space of minimal dimension in which an arbitrary heptagon can be placed. By the formula (1) for the volume of a simplex $A_1A_2A_3A_4A_5A_6A_7$ in E^6 it holds

$$D_6 = \begin{vmatrix} 0 & 1 & 1 & 1 & 1 & 1 & 1 & 1 \\ 1 & 0 & a^2 & b^2 & c^2 & c^2 & b^2 & a^2 \\ 1 & a^2 & 0 & a^2 & b^2 & c^2 & c^2 & b^2 \\ 1 & b^2 & a^2 & 0 & a^2 & b^2 & c^2 & c^2 \\ 1 & c^2 & b^2 & a^2 & 0 & a^2 & b^2 & c^2 \\ 1 & c^2 & c^2 & b^2 & a^2 & 0 & a^2 & b^2 \\ 1 & b^2 & c^2 & c^2 & b^2 & a^2 & 0 & a^2 \\ 1 & a^2 & b^2 & c^2 & c^2 & b^2 & a^2 & 0 \end{vmatrix} = \tag{10}$$

$$= -7(a^6 + 3a^4b^2 - 4a^2b^4 + b^6 - 4a^4c^2 - a^2b^2c^2 + 3b^4c^2 + 3a^2c^4 - 4b^2c^4 + c^6)^2.$$

Now we express the volume of a simplex $A_1A_2A_3A_4A_5A_6$ in E^5 deleting the last row and column in the previous determinant (10). The result is

$$D_5 = 2(2a^4 - 3a^2b^2 + 2b^4 - 3a^2c^2 - 3b^2c^2 + 2c^4)(a^6 + 3a^4b^2 - 4a^2b^4 + b^6 - 4a^4c^2 - a^2b^2c^2 + 3b^4c^2 + 3a^2c^4 - 4b^2c^4 + c^6).$$

The same result we obtain for other 6-tuples of the vertices of a heptagon. The comparison of the determinants D_6 and D_5 gives

$$D_6 = 0 \quad \Rightarrow \quad D_5 = 0. \tag{11}$$

To prove (11) algebraically, we immediately see that the polynomial D_5 is an element of the radical $\sqrt{D_6}$. Note that D_5 *does not* belong to the ideal which is generated by D_6.

Hence, if dimension of a regular heptagon is not six then it is neither five. Therefore the heptagon must lie in the space of dimension four or less.

Suppose that $D_4 = 0$ for all 5-tuples of the vertices of a hepthagon. It is obvious that, see Fig. 4, it suffices to explore three simplices $A_1A_2A_3A_4A_5$, $A_1A_2A_3A_4A_6$, and $A_1A_2A_3A_5A_6$ in E^4. For a simplex $A_1A_2A_3A_4A_5$ we get

$$D_4(12345) = -(a^2 - bc - c^2)(a^2 + bc - c^2)(a^4 - 12a^2b^2 + 8b^4 + 2a^2c^2 - 5b^2c^2 + c^4)$$

and analogously

$$D_4(12346) = -(a^2 - b^2 - ac)(a^2 - b^2 + ac)(a^4 + 2a^2b^2 + b^4 - 5a^2c^2 - 12b^2c^2 + 8c^4),$$

$$D_4(12356) = (ab - b^2 + c^2)(ab + b^2 - c^2)(8a^4 - 5a^2b^2 + b^4 - 12a^2c^2 + 2b^2c^2 + c^4).$$

The condition $D_4 = 0$, i.e. the validity of the equations

$$D_4(12345) = 0, \; D_4(12346) = 0, \; D_4(12356) = 0,$$

means that a heptagon $A_1A_2A_3A_4A_5A_6A_7$ is of dimension three or less.

Investigating the volume of a tetrahedron $A_1A_2A_3A_4$ in E^3 we obtain

$$D_3(1234) = \begin{vmatrix} 0 & 1 & 1 & 1 & 1 \\ 1 & 0 & a^2 & b^2 & c^2 \\ 1 & a^2 & 0 & a^2 & b^2 \\ 1 & b^2 & a^2 & 0 & a^2 \\ 1 & c^2 & b^2 & a^2 & 0 \end{vmatrix} = -2(a^2 - b^2 + ac)(a^2 - b^2 - ac)(a^2 + 2b^2 - c^2).$$

Similarly, considering all quadruples of the vertices of a heptagon, we get another three conditions:

$$D_3(1235) = -2(a^4b^2 - 5a^2b^2c^2 + b^4c^2 + a^2c^4),$$

$$D_3(1245) = -2(a^2 - bc - c^2)(2a^2 - b^2 + c^2)(a^2 + bc - c^2),$$

$$D_3(1246) = -2(a^2 - b^2 - 2c^2)(ab + b^2 - c^2)(ab - b^2 + c^2).$$

We will prove that

$$D_4 = 0 \quad \Rightarrow \quad D_3 = 0, \tag{12}$$

that is, from the hypotheses

$$D_4(12345) = 0, \; D_4(12346) = 0, \; D_4(12356) = 0$$

the conclusions

$$D_3(1234) = 0, \; D_3(1235) = 0, \; D_3(1245) = 0, \; D_3(1246) = 0$$

follow.

Let us denote by $I = (D_4(12345), D_4(12346), D_4(12356))$, and $J = (D_3(1234), D_3(1235), D_3(1245), D_3(1246))$ the respective ideals and compute their radicals. In CoCoA we enter

```
Use R::=Q[abc];
I:=Ideal(-(a^2-bc-c^2)(a^2+bc-c^2)(a^4-12a^2b^2+8b^4+2a^2c^2-
5b^2c^2+c^4),-(a^2-b^2-ac)(a^2-b^2+ac)(a^4+2a^2b^2+b^4-5a^2c^2
-12b^2c^2+8c^4),(ab-b^2+c^2)(ab+b^2-c^2)(8a^4-5a^2b^2+b^4-12a^2c^2
+2b^2c^2+c^4));
Radical(I);
```

and get

$$\sqrt{I} = (a^4 - 2a^2c^2 - b^2c^2 + c^4, a^2b^2 - a^2c^2 - 3b^2c^2 + 2c^4, b^4 - a^2c^2 - 5b^2c^2 + 3c^4). \tag{13}$$

The same result we acquire for $\sqrt{J}$. Hence, the radicals of both ideals I and J are alike.

Thus we proved even more then (12), namely that $D_4 = 0 \Leftrightarrow D_3 = 0$. Whence, if dimension of a regular heptagon is not four then it is neither three. Therefore a heptagon $A_1 A_2 A_3 A_4 A_5 A_6 A_7$ must lie in a plane. We proved the theorem [4], [10]:

Theorem 2. *A regular heptagon lies either in the Euclidean space E^6 or in E^4 or in E^2, i.e. its dimension is either 6 or 4 or 2.*

Remark 3. If we apply the weaker criterion in the theorem above to investigate whether the polynomial $D_3(1234)$ is an element of the ideal I instead of its radical $\sqrt{I}$, we get

```
Use R::=Q[abc];
I:=Ideal(-(a^2-bc-c^2)(a^2+bc-c^2)(a^4-12a^2b^2+8b^4+2a^2c^2-
5b^2c^2+c^4),-(a^2-b^2-ac)(a^2-b^2+ac)(a^4+2a^2b^2+b^4-5a^2c^2
-12b^2c^2+8c^4),(ab-b^2+c^2)(ab+b^2-c^2)(8a^4-5a^2b^2+b^4-12a^2c^2
+2b^2c^2+c^4));
NF(-2(a^2-b^2+ac)(a^2-b^2-ac)(a^2+2b^2-c^2),I);
```

non zero result $-2a^6 + 6a^2b^4 - 4b^6 + 4a^4c^2 + 2b^4c^2 - 2a^2c^4$. Hence the polynomial

$$D_3(1234) = -2(a^2 - b^2 + ac)(a^2 - b^2 - ac)(a^2 + 2b^2 - c^2)$$

is not an element of the ideal I. We again encounter the case in which it is necessary to examine the membership of the conclusion polynomial $D_3(1234)$ in the radical $\sqrt{I}$. A close inspection shows that the polynomial $(D_3(1234))^3$ belongs to the ideal I.

The results above enable us to characterize regular heptagons lying in a plane. A theorem holds:

Theorem 3. *A regular heptagon $A_1 A_2 \ldots A_7$ in E^d with lengths of sides and diagonals $a = |A_i A_{i+1}|$, $b = |A_i A_{i+2}|$, $c = |A_i A_{i+3}|$, where $i = 1, 2, \ldots, n - 1$, is planar if and only if*

$$\begin{aligned}
(a^2 - bc - c^2)(a^2 + bc - c^2) &= 0, \\
(a^2 - b^2 - ac)(a^2 - b^2 + ac) &= 0, \\
(ab - b^2 + c^2)(ab + b^2 - c^2) &= 0.
\end{aligned} \tag{14}$$

Proof. First suppose that a regular heptagon $A_1 A_2 \ldots A_7$ is planar. Denote the equations in (14) successively by $h_1 = 0$, $h_2 = 0$, $h_3 = 0$. Let us prove for instance the first equation $h_1 = 0$. We will show that the polynomial h_1 belongs to the radical $\sqrt{I}$ (13), where $I = (D_4(12345), D_4(12346), D_4(12356))$. Computation in CoCoA gives

```
Use R::=Q[abc];
I:=Ideal(-(a^2-bc-c^2)(a^2+bc-c^2)(a^4-12a^2b^2+8b^4+2a^2c^2-
5b^2c^2+c^4),-(a^2-b^2-ac)(a^2-b^2+ac)(a^4+2a^2b^2+b^4-5a^2c^2
-12b^2c^2+8c^4),(ab-b^2+c^2)(ab+b^2-c^2)(8a^4-5a^2b^2+b^4-12a^2c^2
+2b^2c^2+c^4));
NF((a^2-bc-c^2)(a^2+bc-c^2),Radical(I));
```

the result 0. Similarly we prove other equations in (14).

Conversely, suppose that the equations (14) hold. We are to prove that $D_4 = 0$, that is, $D_4(12345) = D_4(12346) = D_4(12356) = 0$. Let us prove that $D_4(12346) = 0$. We will show that the polynomial $D_4(12346)$ belongs to the ideal $K = (h_1, h_2, h_3)$. The normal form of $D_4(12346)$ with respect to the Gröbner basis of the ideal K

```
Use R::=Q[abc];
K:=Ideal((a^2-bc-c^2)(a^2+bc-c^2),(a^2-b^2-ac)(a^2-b^2+ac),
(ab-b^2+c^2)(ab+b^2-c^2));
NF(-(a^2-b^2-ac)(a^2-b^2+ac)(a^4+2a^2b^2+b^4-5a^2c^2-12b^2c^2
+8c^4),K);
```

equals 0. Similarly we show that $D_4(12345) = 0$ and $D_4(12356) = 0$.

The theorem is proved.

Remark 4. If $K = (h_1, h_2, h_3)$ is the ideal generated by the polynomials on the left sides of (14) then $\sqrt{I} = \sqrt{K}$. Hence, the ideals I and K have the same radicals. From this the Theorem 3 follows.

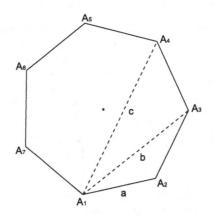

Fig. 4. Convex regular heptagon

Remark 5. To prove that the polynomials h_1, h_2, h_3 on the left sides of (14) belong the ideal I, fails. All these polynomials belong to the radical $\sqrt{I}$. It is easy to show that h_1^4, h_2^4 and h_3^4 belong to the ideal I.

By the Theorem 3 we can compute lengths of sides and diagonals a, b, c of a regular heptagon in E^2. The system of equations

$$a^2 + bc - c^2 = 0, \qquad a^2 - b^2 + ac = 0 \tag{15}$$

characterizes a convex regular heptagon, Fig. 4.

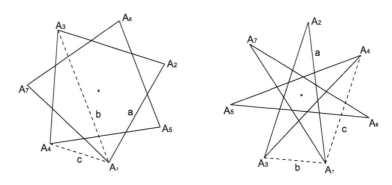

Fig. 5. Non-convex 2-regular and 3-regular heptagons

Similarly, we shall describe the equations for the remaining two non-convex star regular heptagons, Fig. 5. For a heptagon in Fig. 5 on the left (so called 2-regular heptagon) it holds

$$a^2 - bc - c^2 = 0, \qquad ab - b^2 + c^2 = 0,$$

and for a heptagon in Fig. 5 on the right (so called 3-regular heptagon)

$$a^2 - b^2 - ac = 0, \qquad ab + b^2 - c^2 = 0.$$

Regular polygons play an important role in the isoperimetric inequality. The isoperimetric inequality for n-gons in a plane reads [3]:

From all n-gons in a plane of a given perimeter, a regular n-gon has the greatest area.

An analogue of the isoperimetric inequality in a space is as follows [5]:

From all skew n-gons in E^d with the given perimeter find an n-gon of the maximal volume of its convex hull.

This spatial analogue of the isoperimetric inequality for n-gons was successfully solved in spaces of *even* dimension d, where extremal n-gons are more-dimensional analogies of regular polygons [17], [20]. In a space of *odd* dimension d this problem has not been solved yet. A few special cases in E^3 for a skew quadrilateral [16], pentagon and hexagon [13], [14] have been solved. Especially an extremal heptagon in E^3 is of interest since by the theory above it *can not* be regular. Hence a natural question arises, what do extremal n-gons in E^3 for $n \geq 7$ look like?

Acknowledgments

The author is grateful to the anonymous referees for valuable suggestions and recommendations which improved the quality of the text.

References

1. Auric, A.: Question 3867. Intermed. Math. 18, 122 (1911)
2. Berger, M.: Geometry I. Springer, Heidelberg (1987)
3. Blaschke, W.: Kreis und Kugel. Walter de Gruyter & Co, Berlin (1956)
4. Van der Blij, F.: Regular Polygons in Euclidean Space. Lin. Algebra and Appl. 226–228, 345–352 (1995)
5. Bonnesen, T., Fenchel, W.: Theorie der konvexen Körper. Springer, Berlin (1934)
6. Bottema, O.: Pentagons with equal sides and equal angles. Geom. Dedicata 2, 189–191 (1973)
7. Buchberger, B.: Groebner bases: an algorithmic method in polynomial ideal theory. In: Bose, N.-K. (ed.) Multidimensional Systems Theory, pp. 184–232. Reidel, Dordrecht (1985)
8. Chou, S.C.: Mechanical Geometry Theorem Proving. D. Reidel Publishing Company, Dordrecht (1987)
9. Cox, D., Little, J., O'Shea, D.: Ideals, Varieties, and Algorithms, 2nd edn. Springer, Heidelberg (1997)
10. Efremovitch, V.A., Iljjashenko, Ju.S.: Regular polygons. In Russian. Vestnik Mosk. Univ. 17, 18–24 (1962)
11. Grünbaum, B.: Polygons. In: The Geometry of Metric and Linear Spaces. Lecture Notes in Mathematics, vol. 490, pp. 147–184. Springer, Heidelberg (1975)
12. Kapur, D.: A Refutational Approach to Geometry Theorem Proving. Artificial Intelligence Journal 37, 61–93 (1988)
13. Kočandrlová, M.: The Isoperimetric Inequality for a Pentagon in E^3 and Its Generalization in E^n. Čas. pro pěst. mat. 107, 167–174 (1982)
14. Kočandrlová, M.: Isoperimetrische Ungleichung für geschlossene Raumsechsecke. Čas. pro pěst. mat. 108, 248–257 (1983)
15. Lawrence, J.: K-equilateral $(2k + 1)$-gons span only even-dimensional spaces. In: The Geometry of Metric and Linear Spaces. Lecture Notes in Mathematics, vol. 490, pp. 185–186. Springer, Heidelberg (1975)
16. Míšek, B.: O $(n + 1)$-úhelníku v E^n s maximálním objemem konvexního obalu. Čas. pro pěst. mat. 84, 93–103 (1959)
17. Nudel'man, A.A.: Izoperimetričeskije zadači dlja vypuklych oboloček lomanych i krivych v mnogomernych prostranstvach. Mat. Sbornik 96, 294–313 (1975)
18. Recio, T., Sterk, H., Vélez, M.P.: Project 1. Automatic Geometry Theorem Proving. In: Cohen, A., Cuipers, H., Sterk, H. (eds.) Some Tapas of Computer Algebra. Algorithms and Computations in Mathematics, vol. 4, pp. 276–296. Springer, Heidelberg (1998)
19. Recio, T., Vélez, M.P.: Automatic Discovery of Theorems in Elementary Geometry. J. Automat. Reason. 12, 1–22 (1998)
20. Schoenberg, I.J.: An isoperimetric inequality for closed curves convex in even-dimensional Euclidean spaces. Acta Math. 91, 143–164 (1954)
21. Van der Waerden, B.L.: Ein Satz über räumliche Fünfecke. Elem. der Math. 25, 73–96 (1970)
22. Van der Waerden, B.L.: Nachtrag zu "Ein Satz über räumliche Fünfecke". Elem. der Math. 27, 63 (1972)
23. Wang, D.: Gröbner Bases Applied to Geometric Theorem Proving and Discovering. In: Buchberger, B., Winkler, F. (eds.) Gröbner Bases and Applications. Lecture Notes of Computer Algebra, pp. 281–301. Cambridge Univ. Press, Cambridge (1998)

A Maple Package for Automatic Theorem Proving and Discovery in 3D-Geometry

E. Roanes-Macías and E. Roanes-Lozano

Universidad Complutense de Madrid, Facultad de Educación,
Dept. de Algebra, c/ Rector Royo Villanova s/n, 28040-Madrid, Spain
{roanes,eroanes}@mat.ucm.es
http://www.ucm.es/info/secdealg/ERL/

Abstract. A package for investigating problems about configuration theorems in 3D-geometry and performing mechanical theorem proving and discovery is presented. It includes the preparation of the problem, consisting of three processes: defining the geometric objects in the configuration; determining the hypothesis conditions through a point-on-object declaration method; and fixing the thesis conditions. After this preparation, methods based both on Groebner Bases and Wu's method can be applied to prove thesis conditions or to complete hypothesis conditions. Homogeneous coordinates are used in order to treat projective problems (although affine and Euclidean problems can also be treated). A Maple implementation of the method has been developed. It has been used to extend to 3D some classic 2D theorems.

1 Introduction

1.1 Antecedents and State of the Art

We had worked for some time in mechanical theorem proving using algebraic techniques [1,2,3] and we were particularly interested in the cooperation of dynamic geometry systems (DGS) and computer algebra systems (CAS) for this and other purposes.

Some existing DGS use other proving techniques. For instance *Cabry Geometry* [4] is able to try to find counterexamples; *Cinderella* uses a probabilistic method [5,6]; *MathXp* [7,8] uses an automated reasoning engine as prover and a symbolic computation platform as solver...

Regarding the cooperation of DGS and CAS, different solutions have been found:

- to develop a new DGS that is able to communicate with existing CAS (examples: *Geother* [9,10] and its corresponding *Maple* package *Epsilon* [11]; *GDI* [12,13,14])
- to develop a new piece of software that integrates a new DGS and a new CAS that can communicate with each other (examples: *Geometry Expert* [15,16] and *Java Geometry Expert* [17,18]; *Geometry Expressions* [19,20,21]).

F. Botana and T. Recio (Eds.): ADG 2006, LNAI 4869, pp. 171–188, 2007.

(In fact, *Geometry Expressions* can also communicate, bidirectionally (!), with the "external" CAS *Maple* and *Mathematica.*)

We thought an interesting alternative was software reuse [22,23]. Therefore, we developed a connection between *The Geometer's Sketchpad (GSP) v.3* and *v.4* and *Derive* and *Maple*, denoted *paramGeo* [24]. It takes as input either a *GSP v.3 script* or a *GSP v.4 html* description of the construction [25,26,27] and translates it to *Derive* or *Maple* with a translator implemented "ad hoc". Finally, the output of the translator can be manipulated in *Derive* or *Maple* (after loading the corresponding new geometric package). The goal is not only automatic theorem proving and discovery but to input geometric data to the CAS from the DGS for whatever use it is required. Therefore, it leaves the control of the processes to the user.

One of the authors of *GDI* has adopted our idea of software reuse in his new approaches to:

- a connection between the DGS *Cabry Geometry* and *GSP* and the CAS *Mathematica* and *CoCoA* for geometric loci equations finding, denoted *GLI* [28]. It uses the standard *OpenMath* [29] for the communication between the DGS and the CAS
- a connection between the new version of the 3D-DGS *Calques3D* [30] and the CAS *CoCoA* and *Mathematica*, denoted *3D-LD* [31].

We have also studied possible 3D-extensions of classic 2D theorems, some of which we have proven using a CAS (Ceva, Menelaus, Desargues...) [32,34,35,36]. Both the proofs and the implementations have been improved since these works were presented, so we briefly describe then here again.

1.2   paramGeo3D

We have developed in the CAS *Maple* a package denoted *paramGeo3D* that is convenient for introducing 3D geometric data to the CAS *Maple*. It can be used for investigating problems about configuration theorems in 3D-geometry and performing mechanical theorem proving and discovery. It includes the preparation of the problem, consisting of three processes: defining the geometric objects in the configuration; determining the hypothesis conditions through a point-on-object declaration method; and fixing the thesis conditions. After this preparation, methods based both on Groebner Bases and Wu's method can be applied to prove thesis conditions or to complete hypothesis conditions. It extends to 3D the *Maple* package that *paramGeo* (2D) [24] included.

As happens with *paramGeo*, *paramGeo3D* leaves the control to the user and is not only oriented to automatic theorem proving and discovery. Using *Maple* also has the advantage of allowing to use Epsilon's functions [11] (for instance for obtaining a primary ideal decomposition when looking for the components of an algebraic variety).

The main contribution of this work is to provide a convenient environment for algebraic computation in 3D-geometry. The motivation is the lack of such systems (specially if compared to the 2D case).

The main idea underlying the contribution is that an adequate package written in a CAS can be really time saving for doing algebraic computations in 3D-geometry.

The package proves constructive configuration problems.

The article is illustrated with some applications of *paramGeo3D* to mechanical theorem proving and discovery in 3D-geometry (some results are original to the authors of this article).

Maple can plot 2D and 3D geometric objects (after loading the packages *geometry* and *geom3D*, respectively). But, in the 2D case, we did not use *Maple* for plotting and found much more convenient to explore the problem with a dynamic geometry system instead, as mentioned above. Therefore, as the next development of the present work, we plan to connect *paramGeo3D* with *Calques3D*, as done in 2D with *GSP*.

2 Preparations to Prove and Discover

2.1 Defining the Geometric Objects in a Configuration

Some of the objects can be directly defined and others are determined through geometric operations. The directly defined objects are the *initial points* of the configuration. They are defined by their coordinates, numeric or symbolic (the latter considered as parameters).

The other objects are constructed through the adequate concatenation of elementary operations, like: *line* or *segment* determined by two different points; *plane* determined by three non-collinear points; plane *parallel* to another plane through a point; *intersection* of two previously defined objects... (we are treating a constructive geometry).

2.2 Introducing the Hypothesis Conditions

They are declared as membership relations between points and higher dimension geometric objects. To declare $P = (p_0, p_1, p_2, p_3)$ as a point on the object $\phi(x_0, x_1, x_2, x_3) = 0$ is equivalent to impose that the *hypothesis condition* $\phi(p_0, p_1, p_2, p_3) = 0$ is verified. The corresponding *hypothesis polynomial* $\phi(p_0, p_1, p_2, p_3)$ is stored in a list, denoted $LREL$ (List of RELations). The coordinates of the points defined this way are called *variables* and they are stored in another list, denoted VAR. They are easily distinguished (by this method) from *parameters*, that are the non-numeric coordinates of the initial points (those parameters are preserved along all subsequent calculations).

2.3 Fixing the Thesis Conditions

In most configuration problems, the *thesis conditions* are (or can be reduced to) a $P \in \tau$ membership condition (where P is a point and τ is a geometric object) or to a geometric relation among geometric objects in the configuration. In both cases the *thesis polynomials* admits a $\tau(P)$ form.

3   Generating the Automatic Proof

The process of generation of the automatic proof is based on expressing the thesis polynomials as an algebraic linear combination of the hypothesis polynomials. To achieve this, we use methods based on the use of Groebner Bases (GB) [37,38] or pseudo-divisions. They are summarized afterwards.

– In most applications, a simple method consisting of verifying that the GB of the ideal generated by the hypothesis polynomials is equal to the GB of the ideal generated by the hypothesis polynomials together with the thesis polynomials, can be used. It will be denoted *EGB method* (Equal Groebner Bases method).

   This method has an advantage with respect to others: all thesis conditions can be checked at the same time (instead of one by one, what is required by other methods).
– The classic method is based on the radical membership algorithm [39] and will be denoted hereinafter *UGB method* (*Unit Groebner Basis method*).

   This method is very useful when the thesis polynomial is not in the ideal generated by the hypothesis polynomials, but a power of the thesis polynomial is in this ideal.
– A method based on pseudo-divisions is also offered. It essentially consists of Wu's method [40,41,42,43], but adapted to the way hypothesis conditions are declared. (Let us underline that the coordinates of points on objects used for defining hypothesis conditions must be selected in such a way that the number of hypothesis polynomials and the number of variables are equal). This method will be denoted *RSP* (*Reduced Successive Pseudo-divisions*).

   This last method is also applied to automate the process of hypotheses completion, following the ideas introduced by Kapur and Mundy [44] and Recio and Vélez [45].

   Let us give a brief description of the process we use to get new hypothesis conditions, denoted *newHypot*. If the last pseudo-remainder, denoted ρ, obtained when Wu's method is executed, is not zero, then ρ is added to the list *LREL*, and the new variable appearing in the thesis polynomial (and, consequently, in ρ) but not in the list *VAR*, is added to *VAR*. (The process is iterated until the pseudo-remainder vanishes, and the previously calculated ρ polynomials complete the hypotheses).

   It can also be used to determine geometric loci [32]. (In this case, the ρ polynomials obtained prior to the first zero pseudo-remainde determine the locus).

4   Implementation

We have implemented all these processes in *Maple*, in a package that we have called *paramGeo3D* (PARAMetric GEOmetry 3D). It includes several types of commands:

i) commands that define geometric objects, like:
   - `point` given by its coordinates
   - `line` or `segment` determined by two different points

- **plane** determined by three non-collinear points
- **midpoint** of two previously defined points or segment
- **rateOnLine**, that applied to the pair of distinct points (A, B) and a real number r, returns the point P in line AB, such that $\overrightarrow{PB} = r \cdot \overrightarrow{PA}$
- **translate**, that applied to the three points (A, B, C), returns the point P, such that $\overrightarrow{CP} = \overrightarrow{AB}$
- **parallel** plane/line to other previously defined plane/line through a point
- **perpendicular** line/plane to a previously defined plane/line (respectively) through a point
- **dist2** that applied to a point and a point/line/plane, returns the square of the distance between them
- **sphere** determined by its center and a point (or radius) or passing through four points
- **quadric** determined by nine points
- **intersection** of two previously defined objects whose equations have degree 1 or 2 (in case the degree is greater than 2, the intersection can be calculated using the command *PID* of *Epsilon* [11]).

ii) commands that allow to declare memberships:

- **pointOnObject**, that defines hypothesis conditions (when executed, the hypothesis polynomials are automatically added to the list *LREL* and the new variables are added to the list *VAR*)
- in case a **pointOnObject** relation has been already defined for the point, this command is substituted by **pointOnAnotherObject**, that adds the hypothesis polynomials to the list *LREL* but does not add the variables to the list *VAR*.

iii) command **isPlaced**, that, applied to the pair (P, τ) mentioned above, generates the thesis polynomial $\tau(P)$.

iv) command **autProve**, that, applied to the 3-tuple $(P, \tau, \text{method name})$, where the third argument is the abbreviated name of the chosen method (i.e., EGB, UGB, RSP), returns *SUCCESS/FAILURE*, depending on whether $P \in \tau$ has been proved or not.

v) command **newHypot**, that, applied to the pair (P, τ), returns a new hypothesis condition for the thesis condition to hold, if necessary; in other case, the message *No other hypothesis polynomial is necessary* is returned, and the list of algebraic nondegeneracy conditions (i.e., the polynomials that must not vanish to allow to assure that the thesis condition, $P \in \tau$, is true) is stored in the global variable *DEG*.

5 Gallery of Examples

Examples of some of the methods described above are included as illustration afterwards.

5.1 A Simple Example Showing How the Package Works

Example 1. Check if the midpoint of each diagonal of a parallelepiped lies on the other diagonals (see Figure 1).

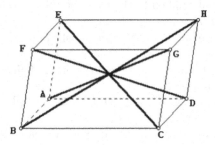

Fig. 1. Concurrency of the diagonals of a parallelepiped

A vertex, A, and its three adjacent ones, B,D,E, can be considered as the initial points of the configuration and the other four vertices are determined from the parallelism of opposite faces. For the sake of simplicity, we can consider A as the coordinate origin, B lying on one coordinate axis and D lying on one coordinate plane:

```
A:=point(1,0,0,0):
B:=point(1,b,0,0):
D:=point(1,d1,d2,0):
E:=point(1,e1,e2,e3):
```

The three plane-faces ABD, ABE, ADE are defined from these vertices. For instance, the first one is defined as follows:

```
ABD:=plane(A,B,D):
```

and the other three plane-faces are defined using parallelism relations:

```
CDG:=parallel(ABE,D):
BCF:=parallel(ADE,B):
EFG:=parallel(ABD,E):
```

We now determine some of the line-edges intersecting faces:

```
BC:=intersection(ABD,BCF):
EH:=intersection(ADE,EFG):
FG:=intersection(BCF,EFG):
```

and the rest of the vertices:

```
C:=intersection(BC,CDG):
G:=intersection(FG,CDG):
H:=intersection(EH,CDG):
F:=intersection(FG,ABE):
```

We now determine the diagonals (as lines):

```
AG:=line(A,G):
BH:=line(B,H):
CE:=line(C,E):
DF:=line(D,F):
```

and the midpoints of the corresponding segments:

```
midAG:=midpoint(A,G):
midBH:=midpoint(B,H):
midCE:=midpoint(C,E):
midDF:=midpoint(D,F):
```

We finally check that the midpoint of each diagonal (for instance, of AG) lies on the other three diagonals:

```
autProve(midAG,BH,RSP);
```
$$SUCCESS$$

```
autProve(midAG,CE,RSP);
```
$$SUCCESS$$

```
autProve(midAG,DF,RSP);
```
$$SUCCESS$$

Meanwhile, the degeneracy conditions are the equations of the form: polynomial in the global variable DEG equal to zero. In this case:

```
DEG;
```
$$[b \cdot e3, -b \cdot d2, -d2 \cdot e3]$$

that are all nonzero if the parallelepiped is non-degenerated.

Note 1. The vertices C, F, G, H could be obtained more briefly using command *translate.*

5.2 An Example Showing How Hypothesis Conditions Are Introduced

Conjecture 1. Let $ABCD$ be a tetrahedron such that D is in the perpendicular plane to line AB through C and such that D is in the perpendicular plane to line BC through A too. Then D is also in the perpendicular plane to line CA through B.

The vertices, A,B,C, of the face ABC of the tetrahedron will be considered as the initial points of the configuration. As the problem is an affine one, we can work with an affine representation, supposing $x_0 = 1$ and the coordinates of these three points will be introduced as lists of three elements. Then our program begins by adding a 1 as the first element (x_0) of the list of coordinates of each one of the three points, in order to perform the computations in homogeneous coordinates in the affine representation:

```
A:=point(0,0,0):
B:=point(1,0,0):
C:=point(c1,c2,0):
```

The three line-sides AB, BC, CA, are defined from these vertices:

```
AB:=line(A,B):
BC:=line(B,C):
CA:=line(C,A):
```

and, denoting XY_Z the plane perpendicular to line XY through point Z, the planes perpendicular to each one of these line-sides through the third vertex of face ABC are defined as follows:

```
AB_C:=perpendicular(AB,C):
BC_A:=perpendicular(BC,A):
CA_B:=perpendicular(CA,B):
```

Now $D = (1, d1, d2, d3)$ is declared as a point on the object AB_C as follows:

```
D:=pointOnObject(1,d1,d2,d3,AB_C):
```

and then the polynomial condition $D \in AB\_C$ is automatically stored in the list $LREL$ and the variables (coordinates) $d1, d2, d3$ are automatically stored in the list VAR. Now, the same point D is declared as a point on the other object BC_A as follows:

```
pointOnAnotherObject(D,BC_A):
```

and then the polynomial condition $D \in BC\_A$ is automatically stored in the list $LREL$ (but the variables variables $d1, d2, d3$ are not stored again in the list VAR, because D is not a new point).

We finally check that the point D lies on the third perpendicular plane CA_B, using any one of the three methods:

```
autProve(D,CA_B,EGB);
```
$$SUCCESS$$

```
autProve(D,CA_B,UGB);
```
$$SUCCESS$$

```
autProve(D,CA_B,RSP);
```
$$SUCCESS$$

When this third method is applied, the degeneracy conditions are stored in the global variable DEG. In this case:

```
DEG;
```
$$[c2^2]$$

that is nonzero if the triangle-face ABC is non-degenerated.

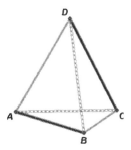

Fig. 2. Orthogonal property of opposite line-sides of a tetrahedron

Thus, we have mechanically proved Conjecture 1. The following corollary follows immediately from this result (because each line included in a perpendicular plane to a given line is also orthogonal to this given line):

Corollary 1. *If two line-sides of a tetrahedron are orthogonal to their opposite ones, then the other two opposite line-sides of the tetrahedron are also orthogonal (see Figure 2).*

(This corollary was mentioned by the anonymous referee).

5.3 Rediscovering Ceva and Menelaus 3D-Theorems Through an Hypothesis Conditions Completion Process

Conjecture 2. Let $ABCD$ be a tetrahedron and let $M \in AB$, $N \in BC$, $P \in CD$ and $Q \in DA$. A condition involving numbers $\frac{MB}{MA}$, $\frac{NC}{NB}$, $\frac{PD}{PC}$ and $\frac{QA}{QD}$, equivalent to the coplanarship of M, N, P and Q, exists (see Figure 3).

Let us rediscover these 3D theorems (that extend to 3D the classic Ceva and Menelaus 2D theorems) through an hypothesis conditions completion process. The process of the proof presented here is shorter than that of [35]. H. Davis recently proved these theorems using synthetic techniques [46].

The initial points A, B, C, D can be defined using command `point`. Then, the hypothesis conditions can be defined using command `rateOnLine`, considering numbers m, n, p, q, such that $\overrightarrow{MB} = m \cdot \overrightarrow{MA}$, $\overrightarrow{NC} = n \cdot \overrightarrow{NB}$, $\overrightarrow{PD} = p \cdot \overrightarrow{PC}$ and $\overrightarrow{QA} = q \cdot \overrightarrow{QD}$. Plane MNP can be defined using command `plane`.

Now, applying command `newHypot` to the pair (Q, MNP), we automatically obtain

$$-1 + m \cdot n \cdot p \cdot q = 0$$

as a necessary condition for $Q \in MNP$.

To verify that it is a sufficient condition, Q is particularized for $q = \frac{1}{m \cdot n \cdot p}$, and applying command `isPlaced` to the pair (Q, MNP), zero is obtained, what confirms that Q belongs to plane MNP.

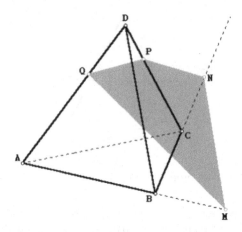

Fig. 3. Extending to 3D Ceva and Menelaus theorems

Thus, we have mechanically discovered the following:

Theorem 1. *Points M, N, P, Q, lying on the oriented consecutive edge-lines AB, BC, CD, DA of tetrahedron $ABCD$ (respectively), are coplanary, if and only if*

$$\frac{MB}{MA} \cdot \frac{NC}{NB} \cdot \frac{PD}{PC} \cdot \frac{QA}{QD} = 1$$

(Let us observe that, as points M, N, P, Q can lie outside the edge-segments, as happens in Figure 3, this result does not only extend Ceva theorem to 3D, but also Menelaus theorem).

5.4 Automatic Proof of Desargues Homological Tetrahedrons Theorem

We suspect that a coplanary condition involving the intersection points of corresponding edge-lines of two homological tetrahedrons (like in Desargues homological triangles theorem), exists. That is, we believe that Conjecture 3 holds.

Conjecture 3. Let $ABCD$ and $A'B'C'D'$ be two tetrahedrons, such that the four lines AA', BB', CC', DD' all meet at a point, O, that is not in any of the faceplanes of $ABCD$ (Figure 4). Then, each one of the six pairs of corresponding edge-lines are intersecting lines and these points of concurrence are coplanary.

This result, that was already proved in [33] using Grassmann algebras, can be automatically proved as follows (the proof is simpler than that of [36]).

The initial points A, B, C, D can be defined using command `point`. Then, the six edge-lines AB, AC, AD, BC, BD, CD and the perspective-lines OA, OB, OC, OD can be defined using the `line` command.

The *hypothesis conditions* consisting of the vertices A', B', C', D' lying on OA, OB, OC, OD, respectively, can be declared using the `pointOnObject` command.

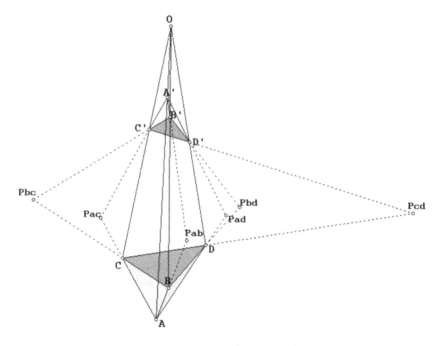

Fig. 4. Configuration of homological tetrahedrons

The face-planes $A'B'C'$, $A'B'D'$, $A'C'D'$, $B'C'D'$ of $A'B'C'D'$ can be defined using the **plane** command. Then, the six edge-lines of $A'B'C'D'$ can be defined using the **intersection** command. Then the intersection of the corresponding edge-lines of both tetrahedrons can be determined applying again command **intersection**. This way six intersecting points are determined:

$$P_{ab} = AB \cap A'B' \ , \ P_{ac} = AC \cap A'C' \ , \ P_{ad} = AD \cap A'D'$$
$$P_{bc} = BC \cap B'C' \ , \ P_{bd} = BD \cap B'D' \ , \ P_{cd} = CD \cap C'D'.$$

The points P_{ab}, P_{ac}, P_{ad} are not collinear and the plane determined by them passes through the other three points of concurrence P_{bc}, P_{bd}, P_{cd}, as can be checked using command **autProve**.

This six intersecting points always verify the incidence conditions shown in Figure 5, as can be checked using command **autProve** again.

Thus, we have mechanically proven Theorem 2.

Theorem 2. *Let $ABCD$ and $A'B'C'D'$ be two tetrahedrons, such that the four lines AA', BB', CC', DD' all meet at a point, O, that is not in any of the face-planes of $ABCD$ (Figure 4). Then, each one of the six pairs of corresponding edge-lines are intersecting lines and these points of concurrence are coplanary and vertices of a complete quadrilateral (Figure 5).*

Note 2. This Desargues 3D theorem is also "self-dual", as the 2D one. That is, interchanging "points" and "lines" gives an essentially equivalent theorem. Let us transcribe it into a version that can be solved using the method above.

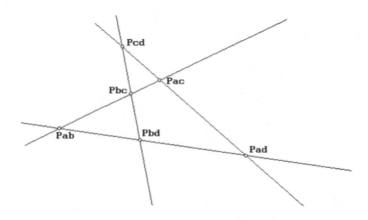

Fig. 5. Configuration of the points of concurrence

Theorem 3. *Let $ABCD$ and $A'B'C'D'$ be two tetrahedrons such that the pairs of corresponding edge-lines AB and $A'B'$, AC and $A'C'$... are intersecting lines and such that their six points of concurrence are vertices of a complete quadrilateral. Then the two tetrahedrons are in perspective position.*

To prove it we begin by fixing a complete quadrilateral and two tetrahedrons with coplanary corresponding edge-lines, whose points of concurrence are the vertices of that complete quadrilateral. We omit this automatic proof for the sake of brevity.

5.5   About the 3D-Extension of Pappus Theorem

We now analyze an extension to 3D of Pappus theorem ("the three intersection points of opposite sides of an hexagon, whose vertices lie alternative on two lines, are collinear"), that is presented here for the first time. Conjecture 4 expresses the natural extension of Pappus theorem to 3D.

Conjecture 4. Given a closed polygonal line with 8 sides, whose vertices lie alternative on two planes and whose opposite sides are secant, then those four points of intersection of opposite sides are coplanary.

Let us denote the planes by α and β, and let us denote by A_1, B_1, A_2, B_2, A_3, B_3, A_4, B_4 the consecutive vertices of the polygonal line in a way such that A_1, A_2, A_3, $A_4 \in \alpha$ and $B_1, B_2, B_3, B_4 \in \beta$ (see Figure 6). Let us denote $L_1 = A_1B_1$, $L_2 = B_1A_2$, $L_3 = A_2B_2,...,L_8 = B_4A_1$ the line-sides of the polygonal line and let us suppose that opposite sides are secant, denoting: $P_1 = L_1 \cap L_5$, $P_2 = L_2 \cap L_6$, $P_3 = L_3 \cap L_7$, $P_4 = L_4 \cap L_8$. Our goal is to check if P_1, P_2, P_3, P_4 are coplanary.

Let us describe a construction that allows to assure that opposite line-sides are secant.

The four first sides of the polygonal line can be chosen arbitrarily. Consequently, the first five vertices, A_1,B_1,A_2,B_2,A_3, can be chosen as the initial points and can be defined using command **point**. Then, plane α, passing through

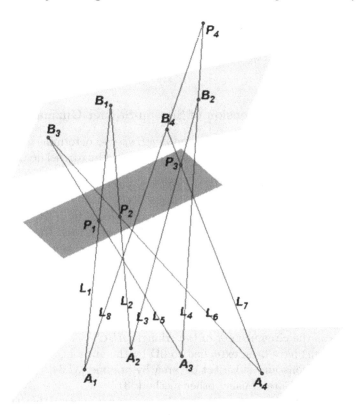

Fig. 6. Configuration of Pappus 3D theorem

A_1, A_2, A_3 (that are supposed to be non-collinear), can be defined using command
plane. Now a new auxiliary point, B_0, can be freely chosen in order to determine
plane β, passing through B_1, B_2, B_0. Command **line** allows now to define the
four first sides $L_1 = A_1 B_1, L_2 = B_1 A_2, L_3 = A_2 B_2, L_4 = B_2 A_3$.

To generate point B_3 we choose $P_1 \in L_1$ (for instance using *rateOnLine*)
and we define L_5 as line $P_1 A_3$ and $B_3 = L_5 \cap \beta$. Point A_4 can be generated
similarly, choosing $P_2 \in L_2$, defining $L_6 = P_2 B_3$ and $A_4 = L_6 \cap \alpha$. Point B_4
can be generated the same way, choosing $P_3 \in L_3$, defining $L_7 = P_3 A_4$ and
$B_4 = L_7 \cap \beta$. The last line of the polygonal is then $L_8 = B_4 A_1$.

Applying command *intersection* to lines L_8, L_4 we can confirm that they are
secant, and we can obtain its common point, P_4.

Finally, *autProof* fails to prove that (in general) P_4 lies on plane $P_1 P_2 P_3$. The
reason is that the result is not true. It is easy to manually find counterexamples.

Summarizing, the conjecture is false. Nevertheless, a related result has been
proven:

Theorem 4. *Given a closed polygonal line with 8 sides, whose vertices lie alternative on two planes and such that their first three line-sides are secant with
their respective opposites, then the fourth side is secant with its opposite.*

Note 3. As L_3 and L_7 are secant, point B_4 is in plane $A_2B_2A_4$ and therefore B_4 must be in the intersection line $r = A_2B_2A_4 \cap \beta$. It can be proved that there exists a point $B \in r$, such that selecting $B_4 = B$, then P_1, P_2, P_3, P_4 are coplanary points.

5.6   About the 3D-Extension of Simson-Steiner-Guzmán Theorem

In order to illustrate the use of command *newHypot* to determine geometric loci, we shall extend to 3D the following result, discovered by Miguel de Guzmán [47]:

> *Given a triangle ABC and three arbitrary projection directions, α, β, γ (not parallel to lines BC, CA, AB, respectively), take an arbitrary point X in the plane of ABC and project it on the lines BC, CA, AB, along directions α, β, γ, obtaining points M, N, P, respectively. Then the locus of all points X such that the area of the triangle MNP is kept unchanged is a conic.*

This result generalizes the well known Simson-Wallace-Steiner theorem. In the latter, the projections are orthogonal to the sides of the triangle ABC (instead of projecting in arbitrary directions) and the locus of point X is a circumference whose center is the circumcenter of the triangle ABC.

Both theorems have been extended to 3D by the authors, substituting triangles by tetrahedrons and substituting area by volume [32,34]. They have also been proved by F. Botana using other method [31].

The results obtained are not exactly the one would expect (that turns out to be false), because the 2D conic or circumference have to be substituted in 3D by a cubic surface instead than by a spherical or quadric surface. These theorems can be proved more easily, using the package described here, as follows.

The vertices O, A, B, C and the locus point X can be defined using command `point`. Then, the four face-planes OAB, OBC, OCA, ABC can be defined using the `plane` command.

The *hypothesis conditions* consisting of the points M, N, P, Q lying on OAB, OBC, OCA, ABC, respectively, can be declared using the `pointOnObject` command.

To define the projection directions, four points D_1, D_2, D_3, D_4 (distinct from O) can be defined using command `point` and then the direction lines OD_1, OD_2, OD_3, OD_4 can be defined using command `line`.

Now, the lines XM, XN, XP, XQ parallel to OD_1, OD_2, OD_3, OD_4 through point X can be defined using command `parallel`.

The *hypothesis conditions* consisting of the points M, N, P, Q lying on XM, XN, XP, XQ, respectively, can be declared using the `pointOnAnotherOb ject` command.

The volume of the tetrahedron $MNPQ$ can be defined as $1/6$ times the determinant of the matrix whose rows are the list of projective coordinates of its vertices. Then, the *thesis condition* is: this volume minus the constant v is equal to zero.

Now, applying command **newHypot** to the thesis condition, we automatically obtain a polynomial expression, h, of degree 3 in the coordinates of X (variables), and whose coefficients are polynomial expressions in the coordinates of the vertices O, A, B, C. Then, the new hypothesis polynomial h is automatically added to the list LREL and one of the coordinates of X is automatically added to the list VAR, as new variable.

Now, applying again command **newHypot** to the thesis condition, the message *No other hypothesis polynomial is necessary* is returned.

Fig. 7. Locus of point X for $v = 0$

Thus, we have mechanically discovered the following theorem (whose additional assertions can be easily verified):

Theorem 5. *The locus of point X, such that $vol(MNPQ) = v$ (v constant) is the cubic surface of equation $h = 0$. The vertices of the tetrahedron $OABC$ are singular points of this cubic surface. In particular, for $v = 0$ (M, N, P, Q coplanary points), the border-lines of the tetrahedron $OABC$ are contained in the cubic surface of the locus (Figure 7).*

6   Conclusions

The package described is convenient for performing investigations and automatic theorem proving and discovery in 3D-geometry. The way the commands are defined allows to automatically determine the equations of geometric objects, hypothesis polynomials, thesis conditions... through an original method of declaring points on objects. The automatic proving algebraic methods have been adapted to the way the hypothesis conditions are declared.

Although projective coordinates are used in order to treat projective problems, it is possible to face affine and Euclidean problems too.

The possibilities of the package are illustrated with a collection of 3D theorems, most of them original to the authors. The *Maple* package, named *paramGeo3D*, is freely available from the authors.

A future extension is the connection with one of the "new" 3D-DGS so that the configuration can be introduced using the computer's mouse, as done in the 2D *paramGeo*. We shall use the new version of *Calques3D*, that, following our request is able to export a readable description of the construction in both *Maple* format. (As far as we know, unlike *Cabri Gèométre II Plus*, *Cabri 3D* does not offer the possibility of returning the text description of the construction).

Acknowledgements

This work was partially supported by the research projects *MTM2004-03175* (Ministerio de Educación y Ciencia, Spain) and *UCM2005-910563* (Comunidad de Madrid - Universidad Complutense de Madrid, research group *ACEIA*).

We would also like to thank the anonymous referees for their most valuable comments.

References

1. Roanes-Macías, E., Roanes-Lozano, E.: Nuevas Tecnologías en Geometría. Editorial Complutense, Madrid (1994)
2. Roanes-Lozano, E., Roanes-Macías, E.: Automatic Theorem Proving in Elementary Geometry with DERIVE 3. The Intl. DERIVE J. 3(2), 67–82 (1996)
3. Roanes-Lozano, E., Roanes-Macías, E.: Mechanical Theorem Proving in Geometry with DERIVE-3. In: Bärzel, B. (ed.) Teaching Mathematics with DERIVE and the TI-92. ZKL-Texte Nr.2, Bonn, pp. 404–419 (1996)
4. http://education.ti.com/educationportal/sites/US/productDetail/us_cabri_geometry.html
5. http://www.cinderella.de
6. Kortenkamp, U.: Foundations of Dynamic Geometry (Ph.D. Thesis). Swiss Fed. Inst. Tech. Zurich, Zurich (1999)
7. http://www.acailab.com/english/mathxp.htm
8. Fu, H., Zeng, Z.: A New Dynamic Geometry Software with a Prover and a Solver. In: Proceedings of VisiT-me'2002. Vienna International Symposium on Integrating Technology into Mathematics Teaching, BK-Teachware, Hagenberg, Austria (2002) (CD-ROM)
9. http://www-calfor.lip6.fr/~wang/GEOTHER/
10. Wang, D.: GEOTHER 1.1: Handling and Proving Geometric Theorems Automatically. In: Winkler, F. (ed.) ADG 2002. LNCS (LNAI), vol. 2930, Springer, Heidelberg (2004)
11. Wang, D.: Epsilon: Software Tool for Polynomial Elimination and Descomposition, http://www-calfor.lip6.fr/~wang/epsilon
12. Botana, F., Valcarce, J.L.: A dynamic-symbolic interface for geometric theorem discovery. Computers and Education 38(1-3), 21–35 (2002)
13. Botana, F., Valcarce, J.L.: A software tool for the investigation of plane loci. Mathematics and Computers in Simulation 61(2), 141–154 (2003)

14. Botana, F., Valcarce, J.L.: Automatic determination of envelopes and other derived curves within a graphic environment. Mathematics and Computers in Simulation 67(1-2), 3–13 (2004)
15. Chou, S.C., Gao, X.S., Zhang, J.Z.: An Introduction to Geometry Expert. In: McRobbie, M.A., Slaney, J.K. (eds.) Automated Deduction - Cade-13. LNCS, vol. 1104, Springer, Heidelberg (1996)
16. http://www.mmrc.iss.ac.cn/gex/
17. Chou, S.C., Gao, X.S., Ye, Z.: Java Geometry Expert. In: Proceedings of the 10th Asian Technology Conference in Mathematics, pp. 78–84 (2005)
18. http://www.cs.wichita.edu/~ye/
19. Todd, P.: A Constraint Based Interactive Symbolic Geometry System. In: Botana, F., Roanes-Lozano, E. (eds.) ADG 2006. Sixth Intl. Workshop on Automated Deduction in Geometry. Extended Abstracts, Universidad de Vigo, Vigo, Spain, pp. 141–143 (2006)
20. Roanes-Lozano, E.: Sobre la colaboración de sistemas de geometría dinámica y de álgebra computacional y el nuevo sistema Geometry Expressions. Bol. Soc. Puig Adam 76, 72–90 (2007)
21. http://www.geometryexpressions.com/
22. Roanes-Lozano, E., Roanes-Macías, E.: How Dinamic Geometry could Complement Computer Algebra Systems (Linking Investigations in Geometry to Automated Theorem Proving). In: Procs. of the Fourth Int. DERIVE/TI-89/TI-92 Conference, BK-Teachware, Hagenberg, Austria (2000) (CD-ROM)
23. Roanes-Lozano, E.: Boosting the Geometrical Possibilities of Dynamic Geometry Systems and Computer Algebra Systems through Cooperation (Plenary Talk). In: Borovcnik, M., Kautschitsch, H. (eds.) Technology in Mathematics Teaching. Procs. of ICTMT-5. Schrifrenreihe Didaktik der Mathematik, öbv&hpt, Wien, Austria, vol. 25, pp. 335–348 (2002)
24. Roanes-Lozano, E., Roanes-Macías, E., Villar, M.: A Bridge Between Dynamic Geometry and Computer Algebra. Mathematical and Computer Modelling 37(9–10), 1005–1028 (2003)
25. Anonymous, The Geometer's Sketchpad User Guide and Reference Manual, vol. 3. Key Curriculum Press, Berkeley, CA (1995)
26. Anonymous, The Geometer's Sketchpad Reference Manual, vol. 4. Key Curriculum Press, Emeryville, CA (2001)
27. http://www.keypress.com/x5521.xml
28. Botana, F., Abanades, M.A., Escribano, J.: Computing Locus Equations for Standard Dynamic Geometry Environments. In: Shi, Y., van Albada, G.D., Dongarra, J., Sloot, P.M.A. (eds.) ICCS – Comput Sci 2007. LNCS, vol. 4488, pp. 227–234. Springer, Heidelberg (2007)
29. http://www.openmath.org
30. http://www.calques3d.org
31. Botana, F.: Automatic Determination of Algebraic Surfaces as Loci of Points. In: Sloot, P.M.A., Abramson, D., Bogdanov, A.V., Gorbachev, Y.E., Dongarra, J.J., Zomaya, A.Y. (eds.) ICCS 2003. LNCS, vol. 2657, Springer, Heidelberg (2003)
32. Roanes-Macías, E., Roanes-Lozano, E.: Automatic determination of geometric loci. 3D-Extension of Simson-Steiner Theorem. In: Campbell, J.A., Roanes-Lozano, E. (eds.) AISC 2000. LNCS (LNAI), vol. 1930, pp. 157–173. Springer, Heidelberg (2001)
33. Hestenes, D., Ziegler, R.: Projective Geometry with Clifford Algebra. Acta Applicandae Mathematicae 23, 25–63 (1991)

34. Roanes-Macías, E., Roanes-Lozano, E.: Extensión a $I\!R^3$ con seudodivisiones de los teoremas de Simson-Steiner-Guzmán. In: Montes, A. (ed.) Actas del 6^0 Encuentro de Algebra Computacional y Aplicaciones EACA 2000, Universitat Politecnica de Catalunya, Barcelona, pp. 331–340 (2000)

35. Roanes-Macías, E., Roanes-Lozano, E.: A Completion of Hypotheses Method for 3D-Geometry. 3D-Extensions of Ceva and Menelaus Theorems. In: Díaz Bánez, J.M., Márquez, A., Portillo, J.R. (eds.) Proceedings of 20th European Workshop on Computational Geometry, Universidad de Sevilla, Seville, pp. 85–88 (2004)

36. Roanes-Macías, E., Roanes-Lozano, E.: A method for outlining 3D-problems in order to study them mechanically. Application to prove the 3D-version of Desargues Theorem. In: González-Vega, L., Recio, T. (eds.) Proceedings of EACA 2004, Univ. of Cantabria, Santander, Spain, pp. 237–242 (2004)

37. Buchberger, B.: An Algorithm for Finding a Basis for the Residue Class Ring of a Zero-Dimensional Polynomial Ideal (in German), Ph.D. Thesis. Math. Ins., Univ. of Innsbruck, Innsbruck, Austria (1965)

38. Buchberger, B.: Applications of Gröbner Bases in non-linear Computational Geometry. In: Rice, J.R. (ed.) Mathematical Aspects of Scientific Software, pp. 59–87. Springer, New York (1988)

39. Cox, D., Little, J., O'Shea, D.: Ideals, Varieties and Algorithms. Springer, New York (2005)

40. Wu, W.T.: On the decision problem and the mechanization of theorem-proving in elementary Geometry. A.M.S. Contemporary Mathematics 29, 213–234 (1984)

41. Wu, W.T.: Some recent advances in Mechanical Theorem-Proving of Geometries. A.M.S. Contemporary Mathematics 29, 235–242 (1984)

42. Wu, W.T.: Mechanical Theorem Proving in Geometries. In: Text and Monographs in Symbolic Computation, Wien, Springer, Heidelberg (1994)

43. Chou, S.C.: Mechanical Geometry Theorem Proving. Reidel, Dordrecht (1988)

44. Kapur, D., Mundy, J.L.: Wu's method and its application to perspective viewing. In: Kapur, D., Mundy, J.L. (eds.) Geometric Reasoning, pp. 15–36. MIT Press, Cambridge MA (1989)

45. Recio, T., Vélez, M.P.: Automatic Discovery of Theorems in Elementary Geometry. Journal of Automated Reasoning 23, 63–82 (1999)

46. Davis, H.: Menelaus and Ceva Theorems and its many applications, http://hamiltonious.virtualave.negt/essays/othe/finalpaper4.htm

47. de Guzmán, M.: An Extension of the Wallace-Simson Theorem: Projecting in Arbitrary Directions. Mathematical Monthly 106, 574–580 (1999)

Geometry Expressions: A Constraint Based Interactive Symbolic Geometry System

Philip Todd

Saltire Software, POB 1565 Beaverton OR,
U.S.A.
philt@saltire.com

Abstract. Real Euclidean geometry is a basic mathematical dialect, not only of high school students, but also of mechanical engineers, graphics programmers, architects, surveyors, machinists, and many more. In this paper, we present "Geometry Expressions": an interactive symbolic geometry package. The aim of the software is to generate algebraic formulas from geometry. It is a further intention of the software that the model should be entered interactively in a style which is convenient to both the geometry consumer groups identified above.

1 Introduction

Interactive Geometry Systems and Computer Algebra Systems both have an established place in education, the former more heavily at the high school level, the latter at the college level. The importance of an interactive symbolic geometry package is that it constitutes a bridge between these two areas of technology: geometry can be entered graphically, symbolic expressions output which may then be transferred to an algebra system for further analysis. It is natural for both consumer groups to merge geometric descriptions with algebraic: the tree that casts a shadow has a height h, and the shadow a length s, a family of mechanical parts is parameterized by exterior and interior diameters D, and d. The author is not aware of any existing software which enables the convenient coexistence of the symbolic with the geometric.

Previous work linking interactive geometry and algebra systems include The Algebraic Geometer, Geometry Expert, paramGeo and Geother and GDI [1-4]. The flavor of these systems is to use a link to an algebra system to prove geometry theorems posed in a dynamical geometry context. Our focus, in contrast, is not on theorem proving per se, but on formula generation. A traditional dynamic geometry system is not the best format for the user interface of a formula generation package, as it does not provide particularly convenient ways to attach symbolic inputs to a model. Quantities are typically derived from locations instead of the other way round. (Distances can, nevertheless, be specified as parameters of translations, angles as parameters of rotations). A constraint based model, however, allows such quantities as distances and angles to be specified directly. This is a much more natural style of user interface for a formula generation package.

F. Botana and T. Recio (Eds.): ADG 2006, LNAI 4869, pp. 189–202, 2007.

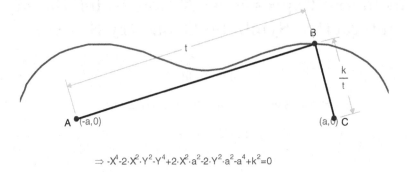

$$\Rightarrow -X^4 - 2 \cdot X^2 \cdot Y^2 - Y^4 + 2 \cdot X^2 \cdot a^2 - 2 \cdot Y^2 \cdot a^2 - a^4 + k^2 = 0$$

Fig. 1. Oval of Cassini defined in terms of a pair of distance constraints

For example, Cassini Ovals are defined as the loci of points the product of whose distance from two fixed points is constant. This is naturally expressed as a pair of symbolic distance constraints from points whose location on a coordinate plane have also been symbolically constrained (Fig. 1). Given these inputs, Geometry Expressions can output the implicit equation of the resulting locus as an expression whose coefficients depend on the undetermined symbols k, the product of the lengths, and a, the absolute value of the x coordinates of the foci.

In this paper, we describe the overall architecture of Geometry Expressions, detail some aspects of the system design, and illustrate through examples the usage of the software.

2   System Architecture

The geometry engine in Geometry Expressions works in the following way [5]

1. A sketch of the geometry along with a set of symbolic constraints is entered by the user.
2. Graph algorithms [6] are used to convert the constraint based description into a sequence of elementary constructions.
3. The construction sequence is executed symbolically resulting in algebraic expressions for the location of each of the geometric objects in the drawing.
4. Measurements made from the drawing are converted into algebraic expressions involving the locations of the geometry objects. The expressions thus obtained are simplified using standard techniques of computer algebra, along with some geometry specific heuristics, and presented to the user.

For example, the drawing of a triangle constrained by two sides and the included angle in (Fig. 2) is converted into this construction sequence:

1. Create a point A at arbitrary location.
2. Create a line AB through A with arbitrary direction.
3. Create a point B on line AB distance a from point A.

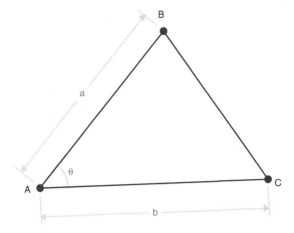

Fig. 2. A geometric figure is specified by a sketch with symbolic constraints (algorithm step 1)

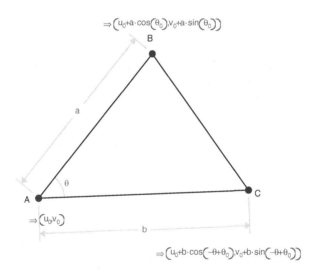

Fig. 3. Geometry Expressions computes symbolic locations for points in a constrained triangle (algorithm step 3)

4. Create a line AC through A and with direction angle θ from line AB.
5. Create a point C on line AC distance b from point A.

Given this construction sequence, Geometry Expressions creates locations for all the points (Fig. 3). Where there is any freedom in the model, Geometry Expressions adds system-generated variables. For example, in fig. 3, the location of A is arbitrary, and the system adds variables u_0 and v_0 as its coordinates. It also adds the variable θ_0 for the arbitrary direction AB.

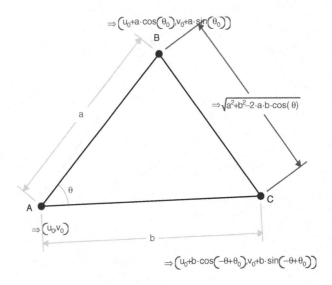

Fig. 4. Output measurements are computed and simplified (algorithm step 4)

When the user asks for a measurement from the drawing, the equivalent algebraic expression is evaluated, simplified and presented. For example, if he asks for the distance between B and C, the distance formula is applied to the symbolic expressions for the coordinates of A and B, simplified and displayed on the diagram (fig. 4).

Within this overall architectural framework, there are a number of design features which are essential to the practicality of the system. We will discuss the following features:

- Intermediate variable retention
- Use of real witness values for variables
- Use of MathML to facilitate two way communication with algebra systems.

2.1   Intermediate Variable Retention

In a purely numerical system, the space and time requirements of the above algorithm are linear in the number of primitive geometric entities. In a symbolic implementation, however, the size of algebraic expressions grows exponentially in the length of the construction sequence. The linear characteristics of the numerical algorithm can be recaptured in the symbolic domain by creating geometric intermediate variables and retaining them in the symbolic representation of the model, and initially in the output measurements.

The user is able to control the display of intermediate variables, by specifying whether they should be retained, and by setting a global granularity parameter. The system substitutes away intermediate variables whose definition is deemed too simple (for example, an intermediate variable which is defined to

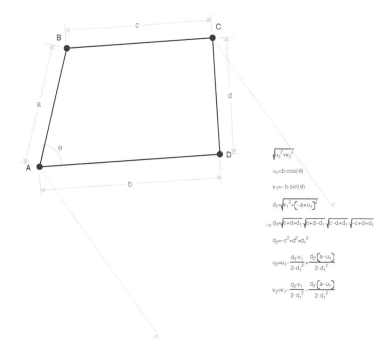

Fig. 5. Distance AD with fine grained intermediate variables. The intermediate variables u_1, v_1, u_2, v_2, d_1, d_2, d_3, are local to the expression and defined, ultimately, in terms of the input variables a,b,c,d, θ.

be a constant would always be considered too simple to retain). The granularity parameter controls the definition of "too simple". Figures 5 and 6 show the same measurement with different settings for the granularity parameter. In Fig. 5, the parameter is set "fine" with the result that more intermediate variables are retained, but their definitions are simple. In Fig. 6, the granularity parameter is set coarser with the result that fewer, but more complicated, intermediate variables are present.

2.2 Witness Values for Variables

The definition of a problem in Geometry Expressions has two components. Algebraically a problem comprises a set of constraints between entities. These constraints correspond to symbolic expressions. In addition, the problem definition contains a sketch of the intended geometry. In general there may be more than one solution to the set of equations corresponding to the constraints. The sketch of the intended geometry is used to choose which solution to use.

In Fig. 7, for example, both triangles ABC and ADC are defined by the same set of constraints (two sides and the non-included angle) however, because ABC is sketched as an acute angle, and ADC is sketched as an obtuse angle,

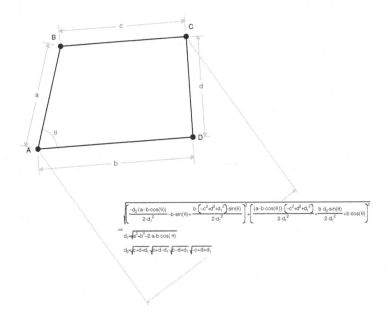

Fig. 6. The same measurement as Fig. 5, but with a coarser setting to the intermediate variable granularity parameter

Geometry Expressions has resolved B and D to different locations - and the symbolic outputs for the lengths BC and DC are indeed different. To reiterate, in terms of symbolic constraints (i.e. algebra) triangles ABC and ADC are defined identically. It is the drawing (i.e. geometry) which leads the symbolic values of BC and DC to differ.

Each solution to the symbolic constraint set represents a family of numerical solutions. Geometry Expressions displays a representative member of the family. In order to do this, it needs to substitute a real number for each of the input variables. For example, in Fig. 7 AB and AD are both specified to have length a. In order to draw a representative solution, the system needs a numeric value for a. In theory, a could have any of a wide range of values, however in practice, the user expects that the representative member of the solution family chosen by the system should be fairly close to his original sketch.

We call the specific numeric value used in the sketch the witness value for the variable. Geometry Expressions has a subsystem for deriving witness values from the sketch, and maintaining witness values throughout its algebra system. Constraints may be specified as expressions involving input variables (Fig. 1, for example). To derive plausible witness values for these inputs, Geometry Expression uses a general purpose numeric root finder. In addition there is a user interface subsystem which allows the user to explicitly set witness values, and to control their behavior on dragging.

A further use of witness values for variables is to allow the automatic creation of assumptions in order to simplify output expressions involving absolute values.

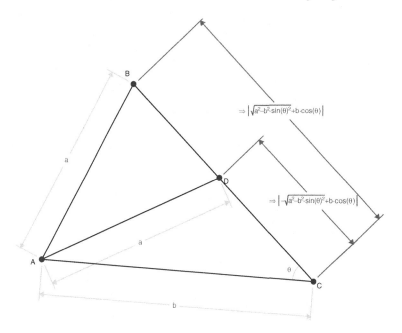

Fig. 7. Triangle ABC is specified in terms of two sides and the non included angle. There are of course two possible such triangles (ADC is the other one). The solution branch which contains the witness triangle is selected by the application.

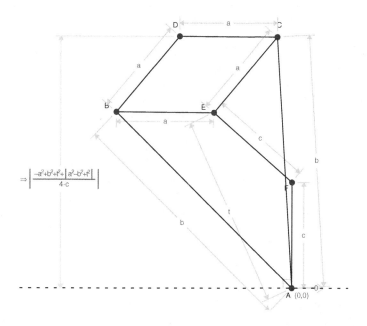

Fig. 8. Paucellier's linkage with the height of D displayed without invoking assumptions based on witness values

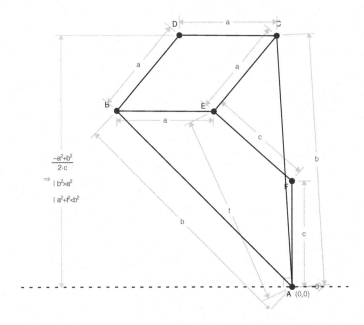

Fig. 9. The height of D simplified with explicit assumptions automatically supplied based on witness values for a,b, and t

Geometry Expressions has both an implicit and an explicit assumptions mechanism. Implicit assumptions are derived from the knowledge that any variables which are used to specify distances or radii must be positive. If the argument of an absolute value can be deduced to be strictly positive or strictly negative based on this information, then the absolute value can be simplified. Such implicit assumptions are applied automatically.

Explicit assumptions may be applied at the user's request. In this case, a real numeric value for the argument of an absolute value is determined based on witness values for any variables which are present. The absolute value is replaced by its argument or by the negative of its argument depending on the sign of this numeric value. An explicit assumption is added to the output expression.

Figures 8 and 9 both show models of Paucellier's linkage. The design of this linkage is such that varying the distance t between B and F should move D in a horizontal straight line. In Fig. 8, the fact that the height of D is independent of t is obscured by the absolute values. In Fig. 9, the addition of assumptions based on witness values for the variables makes its independence of t clear.

2.3   MathML

Geometry Expressions has a special purpose algebra system built in, which is responsible for maintaining and simplifying the expressions generated by the geometric models. However it does not contain a full general purpose CAS. Instead there is a capability for importing and exporting MathML. This allows the user to copy symbolic measurements from Geometry Expressions into the algebra system of his choice, perform some analysis, then, if appropriate, paste the results back into Geometry Expressions.

MathML Example. As an illustration of the use of Geometry Expressions in conjunction with an algebra system, we describe an investigation of the location of the cusps observed in the caustic formed by light from a finite point source reflecting in a cylinder.

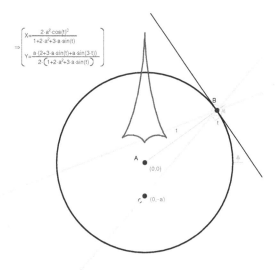

Fig. 10. Parametric equation of the envelope of the rays emanating from the point D (0,-a) and reflected in the unit circle centered at the origin. The parameter t corresponds to the direction AB.

To model this situation in Geometry Expressions (fig 10), we create a circle, AB and an infinite line through B. We constrain the center of the circle to have coordinates (0,0), and constrain its radius to be 1, and we constrain the line through B to be tangential to the circle. We constrain the parametric location of B on the circle to be t. This has the effect of setting the angle between AB and the x axis to be t. We then create a point D and constrain its coordinates to be (0,-a), create a line through D and B, and construct its reflection in the tangent line. Finally we create the envelope of the reflected line as t varies between 0 and 2π.

Two cusps of the envelope curve clearly lie on the y axis and are obtained when B is at parametric locations $\frac{\pi}{2}$ and $-\frac{\pi}{2}$. Creating points on the envelope curve at these parametric locations, we can observe that they do indeed lie at the cusps. Geometry Expressions computes their coordinates.

The parametric location of the other cusps is not so obvious. To compute these, we use the facility of copying and pasting to and from an algebra system

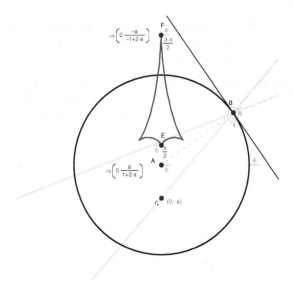

Fig. 11. Coordinates for the points at parametric locations $\frac{\pi}{2}$ and $-\frac{\pi}{2}$ on the envelope curve

(in this case Maple) via MathML. Copying the curve equation into Maple, we differentiate and then solve for the derivatives being simultaneously 0.

$$x := \frac{2\cos(t)^3 a^2}{1 + 2a^2 + 3\sin(t)a} \tag{1}$$

$$y := \frac{1}{2}\frac{(2 + 3\sin(t)a + \sin(3t)a)a}{1 + 2a^2 + 3\sin(t)a} \tag{2}$$

> solve({diff(x,t)=0,diff(y,t)=0},t);

$$\{t = -\frac{\pi}{2}\}, \{t = \frac{\pi}{2}\}, \{t = arctan(-a, RootOf(\_Z^2 - 1 + a^2))\} \tag{3}$$

> allvalues(t = arctan(-a,RootOf(_Z^2-1+a^2)));

$$t = arctan(-a, \sqrt{1 - a^2}), t = arctan(-a, -\sqrt{1 - a^2}) \tag{4}$$

Putting the arctan as the parameter value on a point on the curve, we can have Geometry Expressions give the coordinates of the point (fig 12)

We notice in this example, the use of the two way communication between Geometry Expressions and the algebra system. Curve equations generated from the geometrical problem were exported to the algebra system in order to be differentiated and solved to find cusp locations. Cusp locations were then copied back into Geometry Expressions for further geometrical analysis. The two way communication between these tools makes for a very productive learning and research environment.

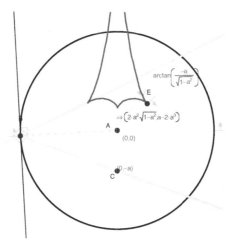

Fig. 12. Coordinates of the cusp, whose parametric location on the envelope curve corresponds to the solution found in Maple

3   Theorem Proving

The interactive nature of the use of Geometry Expressions makes it a useful tool for the discovery as well as the automatic proof of theorems. We illustrate this with an example involving mixtilinear incircles and excircles.

A mixtilinear incircle is tangent to 2 sides of a triangle and (internally) to the circumcircle. A mixtilinear excircle is tangent to 2 sides of a triangle and (externally) to the circumcircle. We show that the ratio of the radii of the mixtilinear excircles and the mixtilinear incircles satisfy an analogous relationship to that between the incircle and excircles.

In Fig. 13, we have constrained a triangle by specifying its side lengths, and created one mixtilinear incircle/excircle pair. Geometry Expressions computes the radii of these circles in terms of the side lengths of the triangle. Observing a degree of commonality between the radii, we display their ratio.

Observing that the ratio displayed has a numerator which is symmetric in a,b,c we are led to consider the sum of the reciprocals of the 3 such ratios. That is the sum of the mixtilinear incircle/excircle radii. Simple algebra leads to the result that if r1, r2, r3 are the radii of the mixtilinear incircles, and if s1, s2, s3 are the radii of the mixtilinear excircles, then:

$$\frac{r_0}{s_0} + \frac{r_1}{s_1} + \frac{r_2}{s_2} = 1 \qquad (5)$$

This is analogous to the relationship between the incircle radius and the excircle radii [7]:

$$\frac{1}{s_0} + \frac{1}{s_1} + \frac{1}{s_2} = \frac{1}{r} \qquad (6)$$

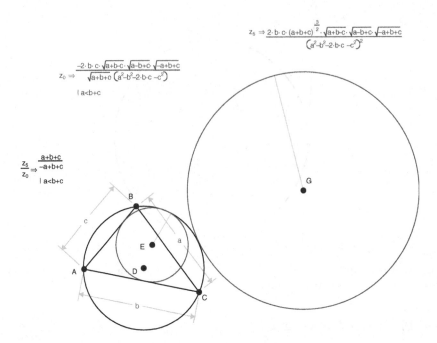

$$z_6 \Rightarrow \frac{2 \cdot b \cdot c \cdot (a+b+c)^{\frac{3}{2}} \cdot \sqrt{a+b-c} \cdot \sqrt{a-b+c} \cdot \sqrt{-a+b+c}}{\left(a^2-b^2-2 \cdot b \cdot c -c^2\right)^2}$$

$$z_0 \Rightarrow \frac{-2 \cdot b \cdot c \cdot \sqrt{a+b-c} \cdot \sqrt{a-b+c} \cdot \sqrt{-a+b+c}}{\sqrt{a+b+c} \left(a^2-b^2-2 \cdot b \cdot c -c^2\right)}$$

$|a<b+c$

$$\frac{z_5}{z_0} \Rightarrow \frac{a+b+c}{-a+b+c}$$

$|a<b+c$

Fig. 13. Expression for the coordinates of the center of the incircle in terms of the coordinates of the triangle vertices

While we make no claims that this result is new in an absolute sense, it was certainly new to us, and its "discovery" facilitated by the formula generation capabilities of Geometry Expressions, working in collaboration with a little human pattern matching.

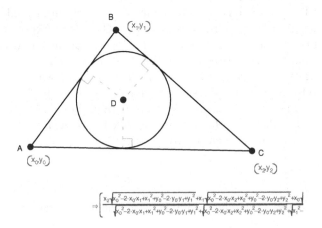

$$\Rightarrow \left[\frac{x_2 \sqrt{x_0^2-2 \cdot x_0 \cdot x_1+x_1^2+y_0^2-2 \cdot y_0 \cdot y_1+y_1^2}+x_1 \sqrt{x_0^2-2 \cdot x_0 \cdot x_2+x_2^2+y_0^2-2 \cdot y_0 \cdot y_2+y_2^2}+x_0 \cdots}{\sqrt{x_0^2-2 \cdot x_0 \cdot x_1+x_1^2+y_0^2-2 \cdot y_0 \cdot y_1+y_1^2}+\sqrt{x_0^2-2 \cdot x_0 \cdot x_2+x_2^2+y_0^2-2 \cdot y_0 \cdot y_2+y_2^2}+\sqrt{x_1^2-\cdots}} \right.$$

Fig. 14. Expression for the coordinates of the center of the incircle in terms of the coordinates of the triangle vertices

4   Further Work

One shortcoming of the approach lies in the fact that the user only has an opportunity to specify symbolic values for an independent set of constraints. In some situations, the form of the symbolic output could be improved by the addition of names for dependent quantities.

For example (Fig. 14), the coordinates of the incenter of a triangle expressed in terms of the vertex coordinates are cumbersome expressions. However, a cursory inspection shows that the terms under the roots are all the distance formula for the lengths of the sides. The complexity of the expression could then be significantly improved if the side lengths are named (Fig. 15).

A topic of further investigation is to extend the basic model so that the user may specify a dependent set of geometric variables.

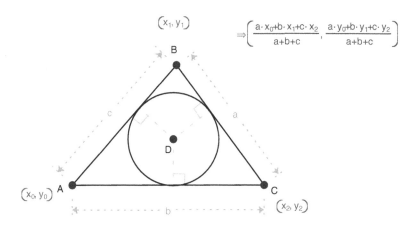

Fig. 15. Expression for the coordinates of the center of the incircle in terms of the coordinates of the triangle vertices and the side lengths

5   Conclusion

A constraint based interactive symbolic geometry system such as Geometry Expressions facilitates a collaborative approach to automated geometry. Collaboration between geometry system and algebra system is enabled by MathML based communication and illustrated by the light caustic example of fig. 10. In that example, the geometry system was used to generate an equation for the caustic curve. The algebra system was used to calculate the location of cusps, and those locations fed back into the geometry system, which was used to display their locus, and to derive the equation of the locus curve.

Collaboration between user and computer is illustrated in the mixtilinear incircle/excircle example of fig.13. The computer is used to generate equations for circle radii in terms of triangle side lengths. Examination and manipulation of these results by a human user leads to the "discovery" of a theorem relating the radii.

References

1. Chou, S.C., Gao, X.S., Ye, Z.: Java Geometry Expert. In: Proceedings of the 10th Asian Technology Conference in Mathematics, pp. 78–84 (2005)
2. Lozano, E.R., Macáas, E.R., Mena, M.V.: A Bridge Between Dynamic Geometry and Computer Algebra. Mathematical and Computer Modelling 37(9-10), 1005–1028 (2003)
3. Lozano, E.R.: Boosting the Geometrical Possibilities of Dynamic Geometry Systems and Computer Algebra Systems Through Cooperation. In: Borovcnik, M., Kautschitsch, H. (eds.) Technology in Mathematics Teaching. Proceedings of ICTMT– 5. öbv & hpt, Schrifrenreihe Didaktik der Mathematik, Viena, vol. 25, pp. 335–348 (2002)
4. Wang, D.: GEOTHER 1.1: Handling and Proving Geometric Theorems Automatically. In: Hong, H., Wang, D. (eds.) ADG 2004. LNCS (LNAI), vol. 3763, pp. 92–110. Springer, Heidelberg (2006)
5. Todd, P., Cherry, G.: Symbolic analysis of planar drawings. In: Gianni, P. (ed.) Symbolic and Algebraic Computation. LNCS, vol. 358, pp. 344–355. Springer, Heidelberg (1989)
6. Todd, P.: A k-tree generalisation that characterises consistency of dimensioned engineering drawings. SIAM Journal of Discrete Mathematics 2, 255–261 (1989)
7. Weisstein, E.W.: "Excircles." From MathWorld–A Wolfram Web Resource, http://mathworld.wolfram.com/Excircles.html

Constructing a Tetrahedron with Prescribed Heights and Widths[*]

Lu Yang[1,2] and Zhenbing Zeng[1]

[1] Institute of Theoretical Computing, East China Normal University,
Beijing 100083, China
lyang@sei.ecnu.edu.cn
[2] Chengdu Institute of Computer Applications, Chinese Academy of Sciences,
Chengdu 610041, China
zbzeng@sei.ecnu.edu.cn

Abstract. Employing a method of distance geometry, we present a symbolic solution to the following problem: express the edge-lengths of a tetrahedron in terms of its heights and widths.

Keywords: generalized Cayley-Menger algebra, widths of a tetrahedron, geometric constraint solving.

1 Introduction

A typical geometric constraint problem requires to find a configuration of points, lines and planes with prescribed pair-wise constraints between these geometric objects. A pair-wise constraint may be the distance or the angle between them. For example, in 5P1L problem, one considers a configuration of 5 points and 1 line in 3-space that is constrained as in Fig. 1. All constraints are distances.

Now, let us discuss an nontypical geometric constraint problem. Given are a set of constraints on four points P_1, P_2, P_3, P_4 in $\mathbf{E}^3$: the distances from each point to the plane determined by the other three points, namely,

$$d(P_1, P_2P_3P_4), \quad d(P_2, P_3P_4P_1), \quad d(P_3, P_4P_1P_2), \quad d(P_4, P_1P_2P_3),$$

and the distances from the line determined by each pair of points to the line determined by the other two, namely,

$$d(P_1P_2, P_3P_4), \quad d(P_1P_3, P_2P_4), \quad d(P_1P_4, P_2P_3).$$

We want to find a realization of the four-point-configuration. In other words, *how to reconstruct a tetrahedron from its heights and widths.* Here P_1, P_2, P_3, P_4 stand for the vertices of a tetrahedron, and h_1, h_2, h_3, h_4 the heights respectively. By *width* we mean the distance between a pair of edges with no intersection. By τ_{ij} denote the width between P_iP_j to its opposite, for $1 \le i < j \le 4$. Clearly that $\tau_{12} = \tau_{34}, \tau_{13} = \tau_{24}, \tau_{14} = \tau_{23}$. So each tetrahedron has three

---

[*] This work is supported in part by NKBRPC-2004CB318003 and NNSFC-10471044.

F. Botana and T. Recio (Eds.): ADG 2006, LNAI 4869, pp. 203–211, 2007.

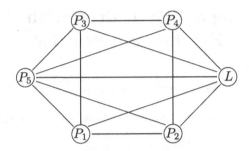

Fig. 1. The $5P1L$ problem: graph vertices represent 5 points and 1 line, while graph edges distances

widths, namely, $\tau_{12}, \tau_{13}, \tau_{14}$. The tetrahedron reconstructing means to express the six edge-lengths $d_{ij} = P_i P_j$ in terms of the heights h_1, h_2, h_3, h_4 and widths $\tau_{12}, \tau_{13}, \tau_{14}$.

Since any tetrahedron has 6 freedoms up to isometries, it is possible to establish a formula connecting the seven quantities $h_1, h_2, h_3, h_4, \tau_{12}, \tau_{13}, \tau_{14}$ The following, which was proved repeatedly in literatures (see [2] and [10] for the history and proofs), gives a very simple equality.

Lemma 1. *The following relation*

$$\frac{1}{h_1^2} + \frac{1}{h_2^2} + \frac{1}{h_3^2} + \frac{1}{h_4^2} = \frac{1}{\tau_{12}^2} + \frac{1}{\tau_{13}^2} + \frac{1}{\tau_{14}^2}$$

holds for any tetrahedron.

So the above problem can also be formulated as: Given any six quantities out of the four heights and three widths of a tetrahedron, can we express the six edge-lengths in terms of the given quantities? This problem has been mentioned by some geometers in different occasions before but still remained unsolved.

In this paper we shall present a solution to this problem by making use of the so-called metric equations and the Jacobi's Theorem about the minors of a matrix. The key is the following formula connecting the dihedral angles, heights and widths of a tetrahedron.

Theorem 1. *Let h_i and h_j be the heights, τ_{ij} the width of a tetrahedron. Draw a triangle ABC with*

$$AB = \frac{1}{h_i}, \quad AC = \frac{1}{h_j}, \quad BC = \frac{1}{\tau_{ij}},$$

then

$$\angle A = \theta_{ij},$$

where θ_{ij} stands for the dihedral angle opposite to the edge $P_i P_j$.

In view of this theorem, we may at first construct a tetrahedron similar to that with given heights and widths, and then compute the similarity constant.

The paper is organized as follows: §2 gives a brief description to the generalized Cayley-Menger Algebra and uses it to derive heights and widths in terms of the edge-lengths (Lemma 2); §3 introduces a formula for dihedral angles (Lemma 3) and proves Theorem 1; §4 shows a method to determine the six edge-lengths of a tetrahedron from its dihedral angles. Lemmas 2 and 3 were established in earlier works of one of the authors of this paper with other collaborators; we shall give a sketch to their proofs and the prerequisite for convenience to readers.

2   Generalized Cayley-Menger Algebra

A very natural attempt to represent the edge-lengths of a tetrahedron with its heights and widths is firstly to establish certain formulas for expressing heights and widths in terms of the edge-lengths, and then solve these equations by taking the edge-lengths as unknowns. It is very easy to compute the heights from the edge-lengths since a height equals three times the quotient of the volume V by the area of a facet, that is,

$$V = \frac{1}{3} \cdot A(P_2 P_3 P_4) \cdot h_1,$$

where $A(P_2 P_3 P_4)$ is the area of triangle $P_2 P_3 P_4$, and both the volume and area can be expressed in terms of edge-lengths by using Cayley-Menger determinants,

$$V^2 = \frac{1}{2^3 \cdot (3!)^2} \cdot D(P_1, P_2, P_3, P_4),$$

$$A(P_2 P_3 P_4)^2 = -\frac{1}{2^2 \cdot (2!)^2} \cdot D(P_2, P_3, P_4),$$

where the Cayley-Menger determinants associated to point sets $\{P_1, P_2, P_3, P_4\}$ and $\{P_2, P_3, P_4\}$ are defined by

$$D(P_1, P_2, P_3, P_4) = \begin{vmatrix} 0 & d_{12}^2 & d_{13}^2 & d_{14}^2 & 1 \\ d_{12}^2 & 0 & d_{23}^2 & d_{24}^2 & 1 \\ d_{13}^2 & d_{23}^2 & 0 & d_{34}^2 & 1 \\ d_{14}^2 & d_{24}^2 & d_{34}^2 & 0 & 1 \\ 1 & 1 & 1 & 1 & 0 \end{vmatrix},$$

and

$$D(P_2, P_3, P_4) = \begin{vmatrix} 0 & d_{23}^2 & g_{24}^2 & 1 \\ d_{23}^2 & 0 & d_{34}^2 & 1 \\ d_{24}^2 & d_{34}^2 & 0 & 1 \\ 1 & 1 & 1 & 0 \end{vmatrix};$$

and the corresponding matrices are called Cayley-Menger matrices associated to $\{P_1, P_2, P_3, P_4\}$ and $\{P_2, P_3, P_4\}$, respectively (see [1,6]).

The formulas for heights can be written in short form as follows. For any $n \times n$ matrix M and integers j, k satisfying $1 \le j, k \le n$, by $M_{j,k}$ denote the j, k

minor of M. Let M and D be the Cayley-Menger matrix and Cayley-Menger determinant associated to the vertices of a tetrahedron $P_1P_2P_3P_4$ respectively, then the heights of the tetrahedron can be expressed in following way.

$$h_1^2 = -\frac{D}{2M_{1,1}}, \quad h_2^2 = -\frac{D}{2M_{2,2}}, \quad h_3^2 = -\frac{D}{2M_{3,3}}, \quad h_4^2 = -\frac{D}{2M_{4,4}}.$$

The Cayley-Menger Algebra was generalized in 1980s-1990s([12,13,14,15,9,11]) to configurations consisting of points, oriented hyperplanes and hyperspheres (shown as in Fig.2) in $\mathbf{E}^n$.

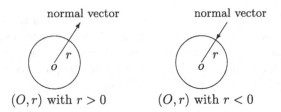

$$(O, r) \text{ with } r > 0 \qquad (O, r) \text{ with } r < 0$$

Fig. 2. oriented hyperspheres in the plane

In the following we quote the generalized Cayley-Menger Algebra for point-plane configurations. Given an n-tuple of points and oriented hyperplanes in an Euclidean space, $\mathcal{P} = (p_1, p_2, \cdots, p_n)$, we define a mapping $g : \mathcal{P} \times \mathcal{P} \to R$ by letting

- $g(p_i, p_j)$ be the square of the distance between p_i and p_j if both are points,
- $g(p_i, p_j)$ be the signed distance from p_i to p_j, if one is a point and the other an oriented hyperplane,
- $g(p_i, p_j)$ be $-\frac{1}{2} \cos(p_i, p_j)$, if both are oriented hyperplanes.

By g_{ij} denote $g(p_i, p_j)$ and G denote the matrix $(g_{ij})_{n \times n}$, and let

$$\delta = (\delta_1, \delta_2, \cdots, \delta_n)$$

where

$$\delta_i = \begin{cases} 1, & \text{if } P_i \text{ is a point,} \\ 0, & \text{if } P_i \text{ is a hyperplane} \end{cases}$$

for $i = 1, \ldots, n$. Then, let

$$M(p_1, p_2, \cdots, p_n) = \begin{pmatrix} G & \delta^T \\ \delta & 0 \end{pmatrix}$$

which is called the Cayley-Menger matrix of $\mathcal{P}$, and let

$$D(p_1, p_2, \cdots, p_n) = \begin{vmatrix} G & \delta^T \\ \delta & 0 \end{vmatrix}$$

which is called the Cayley-Menger determinant of $\mathcal{P}$. The following theorem is an extension to the classical Cayley-Menger determinant.

Theorem 2. *Let $D(p_1, p_2, \cdots, p_n)$ be the Cayley-Menger determinant of an n-tuple of points and oriented hyperplanes in d-dimensional space. If $n \geq d+2$, then*

$$D(p_1, p_2, \cdots, p_n) = 0. \tag{1}$$

By appropriately selecting the sets of the geometric elements, formula (1) creates polynomial equations connecting widths and edge-lengths of a tetrahedron. For instance, construct a plane Π_{12} that passes through points P_3, P_4 and parallels to line $P_1 P_2$, and consider the configuration formed by the following five elements:

$$p_1 = P_1, \ p_2 = P_2, \ p_3 = P_3, \ p_4 = P_4, , \ p_5 = \Pi_{12},$$

as in Figure 3.

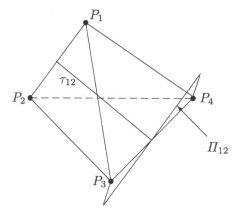

Fig. 3. The plane Π_{12} passes points P_3, P_4 and parallels to line $P_1 P_2$

Following the notations in the last page, for $1 \leq i, j \leq 4$, we have that

$$g_{ij} := g(p_i, p_j) = P_i P_j^2 \qquad \text{and}$$

$$g(p_1, p_5) = g(p_2, p_5) = \tau_{12}, \ g(p_3, p_5) = g(p_4, p_5) = 0, \ g(p_5, p_5) = -\frac{1}{2}.$$

Substituting to the metric equation we get

$$\begin{vmatrix} 0 & g_{12} & g_{13} & g_{14} & \tau_{12} & 1 \\ g_{12} & 0 & g_{23} & g_{24} & \tau_{12} & 1 \\ g_{13} & g_{23} & 0 & g_{34} & 0 & 1 \\ g_{14} & g_{24} & g_{34} & 0 & 0 & 1 \\ \tau_{12} & \tau_{12} & 0 & 0 & -1/2 & 0 \\ 1 & 1 & 1 & 1 & 0 & 0 \end{vmatrix} = 0.$$

This immediately implies the next lemma.

Lemma 2. *Let $P_1P_2P_3P_4$ be a tetrahedron, M and D the Cayley-Menger matrix and Cayley-Menger determinant associated respectively, then*

$$\frac{1}{\tau_{12}^2} = \frac{2M_{1,1} + 2M_{2,2} - 4M_{1,2}}{D}, \quad \frac{1}{\tau_{13}^2} = \frac{2M_{1,1} + 2M_{3,3} + 4M_{1,3}}{D},$$

$$\frac{1}{\tau_{14}^2} = \frac{2M_{1,1} + 2M_{4,4} - 4M_{1,4}}{D}.$$

This result can be found in [10]. Combining this with the height formula it is very easy to prove Lemma 1 by the following computation:

$$(M_{1,1} + M_{2,2} - 2M_{1,2}) + (M_{1,1} + M_{3,3} + 2M_{1,3}) + (M_{1,1} + M_{4,4} - 2M_{1,4})$$
$$= -M_{1,1} - M_{2,2} - M_{3,3} - M_{4,4}.$$

If some six out of the seven quantities, four heights and three widths, are given as constant numbers, we can get the seventh by Lemma 1 and obtain the edge-lengths from height and width formulas by solving a set of equations. But if heights and widths are given in symbols, it is still too complicated to do elimination. Our strategy for this problem is to find certain geometric invariants as intermediate variables that way maybe makes the matter easier. Therefore, the dihedral angles come to mind naturally.

3   Dihedral Angles as Intermediate Variables

In what follows we recall a known result for computing the dihedral angles of a tetrahedron from its edge-lengths.

Lemma 3 (Law of Cosine of Dihedral). *Let $P_11P_2P_3P_4$ be a tetrahedron, M its Cayley-Menger matrix, and θ_{ij} the dihedral angle opposite to edge P_iP_j. Then*

$$\cos(\theta_{ij}) = \frac{-M_{i,j}}{\sqrt{M_{i,i}} \cdot \sqrt{M_{j,j}}}$$

for $1 \leq i < j \leq 4$.

A proof of this lemma can be found in [14]. It involves the application of Theorem 2 combining with the Jacobi's theorem about the minors of matrices (see [7,4,3,8]).

Theorem 3 (Jacobi's Theorem). *For any $n{\times}n$ matrix M and $1 \leq j < k \leq n$, the equality*

$$M_{j,j} \cdot M_{k,k} - M_{j,k}^2 = M_{j,j\,;k,k} \cdot \det(M)$$

holds, where $M_{j,j\,;k,k}$ stands for the determinant of the submatrix of M by deleting j, k-th rows and j, k-th columns .

We just show the sketch of a proof to the Lemma 3. Consider the geometric configuration formed by four vertices of a tetrahedron and two facets $P_2P_3P_4$ and $P_1P_3P_4$. Let

$$p_1 = P_2P_3P_4, \quad p_2 = P_1P_3P_4, \quad p_3 = P_1, \quad p_4 = P_2, \quad p_5 = P_3, \quad p_6 = P_4.$$

and

$$M = \begin{bmatrix} -1/2 & -\cos(\theta_{12})/2 & h_1 & 0 & 0 & 0 & 0 \\ -\cos(\theta_{12})/2 & -1/2 & 0 & h_2 & 0 & 0 & 0 \\ h_1 & 0 & 0 & g_{12} & g_{13} & g_{14} & 1 \\ 0 & h_2 & g_{12} & 0 & g_{23} & g_{24} & 1 \\ 0 & 0 & g_{13} & g_{23} & 0 & g_{34} & 1 \\ 0 & 0 & g_{14} & g_{24} & g_{34} & 0 & 1 \\ 0 & 0 & 1 & 1 & 1 & 1 & 0 \end{bmatrix},$$

the corresponding Cayley-Menger matrix, where θ_{12} is the dihedral angle opposite to edge $P_1 P_2$. Then we have

$$\det(M) = 0, \quad M_{1,1} = 0, \quad M_{2,2} = 0$$

according to Theorem 2, hence $M_{1,2} = 0$ according to Theorem 3. Expanding the determinant $M_{1,2}$, and substituting the height formula for h_1, h_2 into it, we get the law of cosine for the tetrahedron as in Lemma 3.

Now we are ready to prove Theorem 1.

Proof to Theorem 1. Let ABC be a triangle formed by

$$AB = \frac{1}{h_i}, \quad AC = \frac{1}{h_j}, \quad BC = \frac{1}{\tau_{ij}}.$$

Then we have

$$\cos(A) = \left(\frac{1}{h_i^2} + \frac{1}{h_j^2} - \frac{1}{\tau_{ij}^2} \right) \Big/ \left(2 \cdot \frac{1}{h_i} \cdot \frac{1}{h_j} \right).$$

Substitute the formulas for heights and widths into it, we get

$$\cos A = \frac{-M_{i,j}}{\sqrt{M_{i,i}} \cdot \sqrt{M_{j,j}}},$$

and therefore, $\angle A = \theta_{12}$ according to Theorem 3.

4   Determine Edge-Lengths by Dihedral Angles

By means of Theorem 1, we can determine the shape (that is, all the dihedral angles) of a tetrahedron if given the heights and widths. In what follows we show a way to compute edge-lengths via the dihedral angles. The following result ([5]) is the key for this task.

Lemma 4. Let $P_1 P_2 P_3 P_4$ be a tetrahedron, A_i the area of the facet opposite to P_i for $i = 1, 2, 3, 4$, d_{ij} the edge-length of $P_i P_j$, θ_{ij} the dihedral angle opposite to $P_i P_j$, $1 \leq i < j \leq 4$, and V the volume. Then

$$V = \frac{2}{3d_{ij}} A_i A_j \sin(\theta_{ij}).$$

Combining this with Lemma 2,

$$\frac{1}{\tau_{12}{}^2} = \frac{2M_{1,1} + 2M_{2,2} - 4M_{1,2}}{D},$$

and the height formula

$$A_1 = \frac{3V}{h_1}, \quad A_2 = \frac{3V}{h_2}$$

we get the following procedure for computing edge-lengths:

$$g_{ij}^0 = \frac{36}{h_i^2 h_j^2} \sin^2(\theta_{ij}), \quad 1 \le i < j \le 4,$$

$$m = \begin{bmatrix} 0 & g_{12}^0 & g_{13}^0 & g_{14}^0 & 1 \\ g_{12}^0 & 0 & g_{23}^0 & g_{24}^0 & 1 \\ g_{13}^0 & g_{23}^0 & 0 & g_{34}^0 & 1 \\ g_{14}^0 & g_{24}^0 & g_{34}^0 & 0 & 1 \\ 1 & 1 & 1 & 1 & 0 \end{bmatrix},$$

$$V = \sqrt{\frac{2m_{1,1} + 2m_{2,2} - 4m_{1,2}}{\det(m)}} \cdot \tau_{12},$$

$$d_{ij} = \frac{6V}{h_i h_j} \sin(\theta_{ij}), \quad 1 \le i < j \le 4.$$

Now, the tetrahedron has been reconstructed as we asked for.

Acknowledgement

The authors would like to extend their thanks to the anonymous referee for valuable comments and suggestions.

References

1. Blumenthal, L.M.: Theory and Applications of Distance Geometry, Chelsea, New York (1970)
2. Cairns, G., McIntyre, M., Strantzen, J.: Geometric proofs of recent results of Yang Lu. Math. Mag. 66, 263–265 (1993)
3. Gröbner, W.: Matrizenrechnung, Bibliographisches Institut AG, Mannheim, F. R. G, pp. 119–126 (1966)
4. Hodge, W.V., Pedoe, D.: Methods of Algebraic Geometry, Cambridge, vol. I (1953)
5. Lee, J.R.: The Law of Cosines in a Tetrahedron. J. Korea Soc. Math. Ed. Ser. B: Pure Appl. Math. 4, 1–6 (1997)
6. Michelucci, D., Foufou, S.: Using Cayley-Menger Determinants for Geometric Constraint Solving. In: ACM Symposium on Solid Modelling and Applications (2004)
7. Muir, T.: A Treatise on the Theory of Determinants. In: Longmans, W.H. (ed.) revised by. Green, New York, pp. 166–170 (1933)

8. Sippl, M.J., Scheraga, H.A.: Cayley-Menger coordinates. In: Proc. Natl. Acad., USA, vol. 83, pp. 2283–2287 (1986)

9. Yang, L.: Distance coordinates used in geometric constraint solving. In: Winkler, F. (ed.) ADG 2002. LNCS (LNAI), vol. 2930, pp. 216–229. Springer, Heidelberg (2004)

10. Yang, L.: A new method of automated theorem proving. In: Johnson, J.H., Loomes, M.J. (eds.) The Mathematical Revolution Inspired by Computing, pp. 115–126. Oxford University Press, New York (1991)

11. Yang, L.: Solving spatial constraints with global distance coordinate system. International Journal of Computational Geometry & Applications 16(5-6), 533–548 (2006)

12. Yang, L., Zhang, J.Z.: A class of geometric inequalities on finite points. Acta Math. Sinica 23(5), 740–749 (1980)

13. Yang, L., Zhang, J.Z.: The concept of the rank of an abstract distance space. Journal of China University of Science and Technology 10(4), 52–65 (1980) (in Chinese)

14. Yang, L., Zhang, J.Z.: Metric equations in geometry and their applications. I.C.T.P. Research Report, IC/89/281, International Centre for Theoretical Physics, Trieste, Italy (1989)

15. Zhang, J.Z., Yang, L., Yang, X.C.: The realization of elementary configurations in Euclidean space. Science in China A 37(1), 15–26 (1994)

Author Index

Lecture Notes in Artificial Intelligence (LNAI)

Vol. 4659: V. Mařík, V. Vyatkin, A.W. Colombo (Eds.), Holonic and Multi-Agent Systems for Manufacturing. VIII, 456 pages. 2007.

Vol. 4651: F. Azevedo, P. Barahona, F. Fages, F. Rossi (Eds.), Recent Advances in Constraints. VIII, 185 pages. 2007.

Vol. 4648: F. Almeida e Costa, L.M. Rocha, E. Costa, I. Harvey, A. Coutinho (Eds.), Advances in Artificial Life. XVIII, 1215 pages. 2007.

Vol. 4635: B. Kokinov, D.C. Richardson, T.R. Roth-Berghofer, L. Vieu (Eds.), Modeling and Using Context. XIV, 574 pages. 2007.

Vol. 4632: R. Alhajj, H. Gao, X. Li, J. Li, O.R. Zaïane (Eds.), Advanced Data Mining and Applications. XV, 634 pages. 2007.

Vol. 4629: V. Matoušek, P. Mautner (Eds.), Text, Speech and Dialogue. XVII, 663 pages. 2007.

Vol. 4626: R.O. Weber, M.M. Richter (Eds.), Case-Based Reasoning Research and Development. XIII, 534 pages. 2007.

Vol. 4617: V. Torra, Y. Narukawa, Y. Yoshida (Eds.), Modeling Decisions for Artificial Intelligence. XII, 502 pages. 2007.

Vol. 4612: I. Miguel, W. Ruml (Eds.), Abstraction, Reformulation, and Approximation. XI, 418 pages. 2007.

Vol. 4604: U. Priss, S. Polovina, R. Hill (Eds.), Conceptual Structures: Knowledge Architectures for Smart Applications. XII, 514 pages. 2007.

Vol. 4603: F. Pfenning (Ed.), Automated Deduction – CADE-21. XII, 522 pages. 2007.

Vol. 4597: P. Perner (Ed.), Advances in Data Mining. XI, 353 pages. 2007.

Vol. 4594: R. Bellazzi, A. Abu-Hanna, J. Hunter (Eds.), Artificial Intelligence in Medicine. XVI, 509 pages. 2007.

Vol. 4585: M. Kryszkiewicz, J.F. Peters, H. Rybinski, A. Skowron (Eds.), Rough Sets and Intelligent Systems Paradigms. XIX, 836 pages. 2007.

Vol. 4578: F. Masulli, S. Mitra, G. Pasi (Eds.), Applications of Fuzzy Sets Theory. XVIII, 693 pages. 2007.

Vol. 4573: M. Kauers, M. Kerber, R. Miner, W. Windsteiger (Eds.), Towards Mechanized Mathematical Assistants. XIII, 407 pages. 2007.

Vol. 4571: P. Perner (Ed.), Machine Learning and Data Mining in Pattern Recognition. XIV, 913 pages. 2007.

Vol. 4570: H.G. Okuno, M. Ali (Eds.), New Trends in Applied Artificial Intelligence. XXI, 1194 pages. 2007.

Vol. 4565: D.D. Schmorrow, L.M. Reeves (Eds.), Foundations of Augmented Cognition. XIX, 450 pages. 2007.

Vol. 4562: D. Harris (Ed.), Engineering Psychology and Cognitive Ergonomics. XXIII, 879 pages. 2007.

Vol. 4548: N. Olivetti (Ed.), Automated Reasoning with Analytic Tableaux and Related Methods. X, 245 pages. 2007.

Vol. 4539: N.H. Bshouty, C. Gentile (Eds.), Learning Theory. XII, 634 pages. 2007.

Vol. 4529: P. Melin, O. Castillo, L.T. Aguilar, J. Kacprzyk, W. Pedrycz (Eds.), Foundations of Fuzzy Logic and Soft Computing. XIX, 830 pages. 2007.

Vol. 4520: M.V. Butz, O. Sigaud, G. Pezzulo, G. Baldassarre (Eds.), Anticipatory Behavior in Adaptive Learning Systems. X, 379 pages. 2007.

Vol. 4511: C. Conati, K. McCoy, G. Paliouras (Eds.), User Modeling 2007. XVI, 487 pages. 2007.

Vol. 4509: Z. Kobti, D. Wu (Eds.), Advances in Artificial Intelligence. XII, 552 pages. 2007.

Vol. 4496: N.T. Nguyen, A. Grzech, R.J. Howlett, L.C. Jain (Eds.), Agent and Multi-Agent Systems: Technologies and Applications. XXI, 1046 pages. 2007.

Vol. 4483: C. Baral, G. Brewka, J. Schlipf (Eds.), Logic Programming and Nonmonotonic Reasoning. IX, 327 pages. 2007.

Vol. 4482: A. An, J. Stefanowski, S. Ramanna, C.J. Butz, W. Pedrycz, G. Wang (Eds.), Rough Sets, Fuzzy Sets, Data Mining and Granular Computing. XIV, 585 pages. 2007.

Vol. 4481: J. Yao, P. Lingras, W.-Z. Wu, M. Szczuka, N.J. Cercone, D. Ślęzak (Eds.), Rough Sets and Knowledge Technology. XIV, 576 pages. 2007.

Vol. 4476: V. Gorodetsky, C. Zhang, V.A. Skormin, L. Cao (Eds.), Autonomous Intelligent Systems: Multi-Agents and Data Mining. XIII, 323 pages. 2007.

Vol. 4460: S. Aguzzoli, A. Ciabattoni, B. Gerla, C. Manara, V. Marra (Eds.), Algebraic and Proof-theoretic Aspects of Non-classical Logics. VIII, 309 pages. 2007.

Vol. 4457: G.M.P. O'Hare, A. Ricci, M.J. O'Grady, O. Dikenelli (Eds.), Engineering Societies in the Agents World VII. XI, 401 pages. 2007.

Vol. 4456: Y. Wang, Y.-m. Cheung, H. Liu (Eds.), Computational Intelligence and Security. XXIII, 1118 pages. 2007.

Vol. 4455: S. Muggleton, R. Otero, A. Tamaddoni-Nezhad (Eds.), Inductive Logic Programming. XII, 456 pages. 2007.

Vol. 4452: M. Fasli, O. Shehory (Eds.), Agent-Mediated Electronic Commerce. VIII, 249 pages. 2007.

Vol. 4451: T.S. Huang, A. Nijholt, M. Pantic, A. Pentland (Eds.), Artificial Intelligence for Human Computing. XVI, 359 pages. 2007.

Vol. 4442: L. Antunes, K. Takadama (Eds.), Multi-Agent-Based Simulation VII. X, 189 pages. 2007.

Vol. 4441: C. Müller (Ed.), Speaker Classification II. X, 309 pages. 2007.

Vol. 4438: L. Maicher, A. Sigel, L.M. Garshol (Eds.), Leveraging the Semantics of Topic Maps. X, 257 pages. 2007.

Vol. 4434: G. Lakemeyer, E. Sklar, D.G. Sorrenti, T. Takahashi (Eds.), RoboCup 2006: Robot Soccer World Cup X. XIII, 566 pages. 2007.

Vol. 4429: R. Lu, J.H. Siekmann, C. Ullrich (Eds.), Cognitive Systems. X, 161 pages. 2007.

Vol. 4428: S. Edelkamp, A. Lomuscio (Eds.), Model Checking and Artificial Intelligence. IX, 185 pages. 2007.